Amel Abbad
Samir Bentata
Fodil Hamzaoui

Optoelectronic and Magnetic Properties of Inorganic Materials

Amel Abbad
Samir Bentata
Fodil Hamzaoui

Optoelectronic and Magnetic Properties of Inorganic Materials

Ab initio theoretical calculations

LAP LAMBERT Academic Publishing

Imprint

Any brand names and product names mentioned in this book are subject to trademark, brand or patent protection and are trademarks or registered trademarks of their respective holders. The use of brand names, product names, common names, trade names, product descriptions etc. even without a particular marking in this work is in no way to be construed to mean that such names may be regarded as unrestricted in respect of trademark and brand protection legislation and could thus be used by anyone.

Cover image: www.ingimage.com

Publisher:
LAP LAMBERT Academic Publishing
is a trademark of
Dodo Books Indian Ocean Ltd. and OmniScriptum S.R.L publishing group

120 High Road, East Finchley, London, N2 9ED, United Kingdom
Str. Armeneasca 28/1, office 1, Chisinau MD-2012, Republic of Moldova, Europe
Managing Directors: Ieva Konstantinova, Victoria Ursu
info@omniscriptum.com

Printed at: see last page
ISBN: 978-3-659-74239-2

Optoelectronic and Magnetic Properties of Inorganic Materials
Ab initio theoretical calculations

Amel ABBAD, Wissam BENSTAALI, Samir BENTATA & Fodil HAMZAOUI

SUMMARY

Introduction

The search for materials with optimized optical, structural and magnetic properties has attracted much attention in this last decade. Using the charge and spin degrees of freedom of electron in a single device are the main objective of the spintronics field. The coupling between the charge carriers in a semiconductor and the electron spins of a ferromagnetic metal doped into the semiconductor can be used for many magneto/spin-electronic devices. The ultimate success of these materials for applications in "spintronics" relies on identifying the best candidates for this reason.

In this workbook we present a selection of our previous works done on a variety of inorganic compounds which have been extensively studied because of their promising magnetic and optoelectronic applications.

The calculations of the magnetic and the optical properties presented in this work have been performed within the framework of density functional theory (DFT) using the full-potential linearized augmented plane wave method (FPLAPW) as implemented in the WIEN2K code. The electrons exchange-correlation energy was described in the local spin density approximation (LSDA) and the generalized gradient approximation (GGA). The muffin-tin (MT) radii of the investigated atoms were chosen from literature. Basis functions were expanded as combinations of spherical harmonic functions inside non-overlapping spheres around the atomic sites (MT spheres) and in plane waves in the interstitial region.

The Brillouin first zone of the investigated compound is chosen regarding the nature of its structure and after performing an energy convergence test. The wave functions in the interstitial region were expanded in plane waves with a cutoff of $k_{max}=8/R_{MT}$ (where R_{MT} is the average radius of the MT spheres). For a self-consistent field, the convergence for the total energy was set to 10^{-4} Ry. The muffin-tin radius R_{MT} is based on two conditions: (i) no core charge leaks out of MT spheres, (ii) no overlapping is permitted between spheres. The structural optimization was obtained by fitting the total energy to Murnaghan's equation of state.

The optical properties were calculated using the Optic code implemented in Wien2k. Recognizing that the LSDA underestimates the band gap, we took into account in our optical calculations, for each studied structure, a correction of the band gap which corresponds to the difference between the measured gap and the calculated one.

Our reported calculations highlight some precisions about the found experimental results and offer a large theoretical prediction of the magnetic and the optoelectronic properties of the investigated compounds.

Keywords:
Total Density of States, Magnetic Moments, Optical Properties, First-Principles Calculations, Drude Model.

Influence of Ni-Ni separation on the optoelectronic and magnetic properties of Ni-doped Cubic Cadmium Sulphide

W.Benstaali[1], S. Bentata[1], H.A. Bentounes[2], A.Abbad[2] and B.Bouadjemi[1]

[1]Laboratory of Technology and of Solids Properties
Faculty of Sciences and Technology, BP227
Abdelhamid Ibn Badis University, Mostaganem (27000) Algeria
[2] Signals and Systems Laboratory (LSS)
Faculty of Sciences and Technology, BP227
Abdelhamid Ibn Badis University, Mostaganem (27000) Algeria

*E-mail: ben_wissam@yahoo.fr

Abstract

The full-potential linearized augmented plane wave method (FP-LAPW) within the local spin density approximation (LSDA) is used to calculate the electronic, magnetic and optical properties of Ni-doped CdS. The results show that NiCdS for a concentration of 6.25% behaves as a semiconductor and become half-metallic and ferromagnetic when Ni concentration is increased to 12.5%. Furthermore, the magnetic coupling between the Ni atoms is sensitive to the Ni-Ni distance. The rising of the separation of the two Ni atoms leads to a significant decrease in magnetic coupling between them and an amelioration of optical properties.
Keywords: magnetic moment, half-metallic, Configuration, optical properties.

1. Introduction:

The synthesis of binary metal chalcogenides of group II–VI semiconductors has been the center of recent scientific research due to their crucial luminescent properties, non-linear optical properties, quantum size effect and other important physical and chemical properties [1]. Cadmium sulfide (CdS) is one of the most important II–VI group semiconductors. It has attracted increased attention in recent years because of its wide direct band gap energy (2.5 eV) [2], its optical and electrical properties, and stability. Pure CdS is suitable for application as a window layer in solar cells, optoelectronics and photocatalysts [3–7] and also when it is doped with Indium [8]. In recent years, Transition Metal (TM) doped nanoparticles, have fascinated wide scientific attention because of their unique optical properties [9-14] and their potentiality for various applications other than biomedical labeling [15, 16]. Lastly, the efficiency of CdS semiconductor was improved by changing its optical and/or electrical properties by doping with some transition metals such as Copper [17,18], Manganese [19,20] and Cobalt [21]. Experimentally, B. Srinivasa et al [22], have successfully prepared Ni doped CdS nanoparticles by conventional chemical co-precipitation method and found that doping CdS with Ni

4

modify the luminescence properties by creating shallow acceptor states. However, no theoretical calculations on Ni-doped CdS are available to our knowledge. Consequently, in this letter we report a theoretical study of the magnetic, electronic and optical properties of Nickel substituting in cubic CdS semiconductor. The magnetic interaction between the Ni atoms and the optical properties were found sensitive to the Ni-Ni separation in the supercell.

2. Calculation:

The calculations presented in this work have been performed within the framework of density functional theory (DFT) [23, 24] using the full-potential linearized augmented plane wave method (FPLAPW) [25] as implemented in the WIEN2K code [26]. The electrons exchange-correlation energy was described in the local spin density approximation (LSDA) [27, 28]. The muffin-tin (MT) radii of Cd, S, and Ni were chosen to be 2.30, 1.70 and 1.95 respectively. Basis functions were expanded as combinations of spherical harmonic functions inside non-overlapping spheres around the atomic sites (MT spheres) and in plane waves in the interstitial region. We have used 150 k points in the first Brillouin zone (68 special k points in the reduce wedge corresponding to a mesh of (4x4x8)), this value of k point was obtained after an energy convergence test. The wave functions in the interstitial region were expanded in plane waves with a cutoff of $k_{max}=8/R_{MT}$ (where R_{MT} is the average radius of the MT spheres). For a self-consistent field, the convergence for the total energy was set to 10^{-4} Ry. The muffin-tin radius R_{MT} is based on two conditions: (i) no core charge leaks out of MT spheres, (ii) no overlapping is permitted between spheres. The zinc blend structure of CdS was used [29-31]. The supercell employed contains 32 atoms, which corresponds to a 2×2×1 supercell of CdS. The calculated equilibrium lattice parameter which minimizes the total energy as a function of the cell volume and which relaxes the atomic positions was found (a=5.7640 Å). This value of (a) is in good agreement with experimental ones [32]. It was obtained by fitting the total energy to Murnaghan's equation of state [33] for structural optimization. The optical properties were calculated using the Optic code implemented in Wien2k. The number of k points used for optical properties was 400. Recognizing that the LSDA underestimates the band gap (see fig (1-a) later), we took into account in our optical calculations, for each studied structure, a correction of the band gap which corresponds to the difference between the measured gap (2.42eV) and the calculated one.

3. Results and discussions

3.1 Magnetic and Electronic properties:

Total density of state (DOS) of CdS and Ni-doped CdS are shown in Fig. 1. From Fig 1(a), we can see that the Fermi level is located in the gap region and that pure CdS behaves like a semiconductor. In fact, the DOS curves for spin-up and spin-down states are totally symmetric: consequently there is no net magnetic moment in this supercell. Valence band (VB) in pure CdS is mainly composed by two parts, top valence band (-3.36 to -0.25 eV) and bottom valence band (-5 to -3.49 eV), as illustrated in Fig. 2(a). It is worth noting that the valence band is mainly contributed by S (p) with a little contribution of Cd (d) orbitals. Fig 1(b) and 2(b) show the total and partial DOS of NiCdS. In this case, the Fermi level passes also through the band-gap which means that the compound keeps its semi conducting behaviour, we can also observe a reduction in the band gap compared to pure CdS, this is due to the appearance of localised states in the minority spin of the valence band.

To further explore the effect of Ni concentration on the magnetic and electronic properties of NiCdS system, we have performed additional calculations where two Cd atoms were replaced by two Ni atoms in the supercell, corresponding to a higher Ni concentration of 12.5%. In figure 1(c), we plot the total density of states of $Ni_2Cd_{14}S_{16}$. The characteristic feature of the DOS is the deep Ni localized states in the semiconducting gap of CdS. In this case, we note that the Fermi level (EF) is pinned in the valence band, which means that Ni-doped CdS is of p-type conductivity character. The minority spin channel (spin-down) is fully occupied, while the majority spin channel is empty resulting in a half metallic ferromagnet with 100% spin polarization. There is also a visible overlap between Ni (d) and S (p) states (Fig.2 (c))

To see if Ni-Ni distance has an impact on the electronic and magnetic properties of Ni doped CdS, we have studied three possible configurations in which the two Ni atoms replace Cd at the nearest, second-nearest, and third-nearest sites in the supercell (Fig. 03) . In configuration (I), the distance d_{Ni-Ni} between the two Ni atoms is 4.07 Å, in configuration (II), it is 5.75 Å and in configuration (III), it is 8.14 Å. The corresponding DOS's are plotted on figure. 04. First of all, we observe a half-metallic behavior for the three configurations, in the sense that the Fermi level state density is finite for the minority spin and is zero for the majority spin. Secondly, from the partial DOS's shown on figure (5), we can note that when the Ni-Ni separation decreases, the Ni (d) states become more delocalized.

Concerning the p-type character of Ni-doped CdS mentioned above, we can conclude that the large PDOS values of Ni and its delocalized nature in configuration (I) are considered to be the reasons of good conductivity of NiCdS, and Ni (d) states contribute excess holes as major carriers.

Total energies corresponding to both FM and AFM spin alignments were calculated to determine the preferred magnetic ground state for the three configurations. The Ni-Ni distance and energy difference between AFM and FM states ($\Delta E = E_{AFM} - E_{FM}$) for each configuration are listed in Table 01. It was found that configuration (I), with FM coupling, has the highest value of ΔE comparing to the two other configurations, and it is obvious that the energy difference ΔE decreases as the Ni-Ni distance increases. Configuration (I) has a stable FM state (42 meV) which is lower in energy than the AFM state, we can say that the two Ni atoms prefer to occupy near sites, and cluster around S atoms. The FM coupling between the Ni atoms is mediated by the S atom as it can be seen from the overlap in the DOS between the S (p) and the Ni (d) states, especially occurring in the minority orbitals as shown in Figure 5.

The magnetic study shows that the magnetization of $Cd_{14}Ni_2S_{16}$ is 4 μ_B per supercell with a localized magnetic moment of 1.26 μ_B per Ni atom. The nearest neighbor host atoms are weakly polarized (induced moments of +0.1 μ_B on S sites and 0.01 μ_B on Cd sites). It should be noted that the values of the magnetic moments vary slightly for the three configurations.

3.2 Optical properties:

When a material is in the presence of an external electric field, its polarization response is measured through the dielectric function. From the knowledge of the real and imaginary parts of the dielectric tensor, it is possible to predict other important optical properties such as the absorption coefficient, the refractive index and the reflectivity [34-38].

Since absorption coefficient and dielectric function depend strongly on the transitions between valence-band and conduction-band (CB), the understimated LSDA band-gap energies need to be corrected, so we have calculated the optical properties using the scissor operator. The scissor operator applied, accounts the difference between the experimental band gap and the calculated one of each configuration. The computed results for the real and imaginary parts of the dielectric function $\varepsilon\ (\omega)$ of pure CdS and Ni-doped CdS for the three different configurations are presented in Fig. 6. The imaginary part ε_2 for pure CdS starts at 2.43 eV, which corresponds to the band gap. This figure shows the appearance of a significant peak, at low energies in the band gap, for the doped structures, especially for configuration (III) in both real and imaginary parts, in contrast to pure CdS where no bumps are seen. These peaks situated in the infrared region are mainly due to the free electrons and holes generated by the doping process. This effect is called "free carrier absorption". Given that Ni-Ni distance is very high in configuration (III), the important peak seen is probably due to the d-d interband transitions between the two Ni atoms. We also note that for each configuration, these peaks have different values and are red-shifted when Ni-Ni distance decreases. The band-edge after Ni-doping is red-shifted which clearly indicates that Ni^{2+} ions were really incorporated into the CdS lattice [39].

The real part of the dielectric function for the doped structures are negative in a very small region in the infrared, in opposite to the pure structure. The negative values of $\varepsilon_1\ (\omega)$ show that the incident electromagnetic waves, in this energy region, are entirely reflected, consequently, the material shows metallic nature.

Since the dielectric function is used to calculate the absorption coefficient, the polarization response similarities are taken into account in absorption (Fig 7). The figure reveals several prominent peaks above absorption edge for pure CdS (E_g=2.43eV), which are due to excitonic effects. These appeared peaks become broad and are less pronounced after the doping process. They are due to interband transitions of TM impurities from the topmost VB to the lowest CB (Fig. 4). For each configuration, there are several peaks in the infrared, because it is a p-type semiconductor. P-type semiconductors show an additional absorption mechanism in low energies due to intervalence band transitions [38]. It is worth noting that the peak which corresponds to maximum absorption coefficient increases after Ni-doping and that the curves are blue-shifted. These results are in a good agreement with experimental ones [40]. Furthermore, we can see an important peak in the absorption coefficient centred at 2.95eV. This peak explains the important infrared absorption of maximum light at this wave length. The three different configurations have significant absorption in large range of photon energies, and consequently, the results obtained will allow a better design of thin films solar cells which use NiCdS as the absorber layer in the visible and ultraviolet parts of the spectrum, they can also be used as a good sensor in the infrared especially when the impurities in the structure are largely separated.

In Fig. 8, we plot the different reflectivities $R(\omega)$ for the pure and Ni-doped CdS. The peaks at 1.83eV, 2.32eV and 3.19eV for configuration (I), configuration (II) and configuration (III) respectively, are clearly related to the peaks seen in the dielectric function. For configuration (III), we can note that the reflectivity is very close to 80 % at low energies. In this region, pure CdS has low reflectivity (20%) and reaches its maximum value (58%) for 10.1eV. Moreover, the reflectivities for the three compounds are significantly enhanced in the energy range between 15 eV and 20 eV as a result of interband transitions while both real part of the dielectric function $\varepsilon_1(\omega)$ and imaginary part $\varepsilon_2(\omega)$ are close to zero in this energy range (see Fig. 6).

A further attractive characteristic of figure 8 is the shift of the curves towards higher energies after the doping process. The doped materials have a larger reflectivity energy range than the undoped ones, in particular in configuration (III) which has the greatest value of the width of the reflectivity.

Knowing that CdS is extremely used in optoelectronics, we have studied these optical properties for different configurations in order to find the corresponding structure which gives the best properties according to the desired application.

4. Conclusion:

In this work, we used the density functional theory in the spin local density approximation in order to study the electronic, magnetic and optical properties of Ni doped CdS. We have studied three possible configurations in which the two Ni atoms replace Cd at the nearest, second-nearest, and third-nearest sites in the supercell. Our results show that CdS doped with 6.25% of nickel is

9

an intrinsic semiconductor, but when it is doped with 12.5% of nickel, it becomes half-metallic and ferromagnetic material. When the two Ni atoms are nearest-neighbours, they showed delocalized quality and might be promising half-metallic and ferromagnetic materials for applications in spintronics. Additionally, the optical properties of pure CdS and Ni-doped CdS for the three configurations were also calculated. We can note the appearance of a significant peak for the doped structures, at low energies, especially for configuration (III) in all the optical curves. This is probably due to the large separation between the two Ni atoms. So, NiCdS can be a good candidate in the design of thin film solar cells in the visible and ultraviolet parts of the spectrum, and a good sensor in the infrared especially when the impurities in the structure are largely separated. As a result, we can deduce from this study that the increase of Ni-Ni separation in the supercell, leads to a reduction in the magnetic coupling and an improvement of optical properties. As perspective to this work, we can study the effect of temperature on the ferromagnetism of the three configurations.

Acknowledgment

The authors thank Pr. Bouhafs and Pr. Abbar for their helpful discussions.

References:

[1] A.V. Murugan, R.S. Sonawane, B.B. Kale, S.K. Apte, A.V. Kulkarni, Mater. Chem. Phys. 71 (2001) 98–102.

[2] A. Phuruangrat, T. Thongtem, S. Thongtem, Mater. Lett. 63 (2009) 1538–1541

[3] Zhang, H. Ma, X. Ji, Y. Xu, J. Yang, D. Chem. Phys. Lett., 377, (2003) 654–657

[4] J. Palm, V. Probst, W. Stetter, R. Toelle, S. Visbeck, H. Calwer, T. Niesen, H. Vogt, O. Hernandez, M. Wendl, and F. H. Karg: Thin Solid Films 544 (2004) 451–452

[5] Schropp, R.E.I.; Zeman, M.Modelling, Materials and Device Technology; Springer: Berlin, Germany, 1998.

[6] A. Phuruangrat, T. Thongtem, S. Thongtem, Mater. Lett. 63 (2009) 1538–1541

[7] N.H. Kim, S.H. Ryu, H.S. Noh, W.S. Lee, Material Science in Semiconductor Processing 15 (2) (2012) 125–130

[8] F.Atay, V. Bilgin, I. Akyuz, S. Kose. *Materials Science in Semiconductor Processing 6 (4) (2003) 197-203*

[9] O.E. Raola, G.F. Strouse, Nano Lett. 2 (2002) 1443-1447

[10] F.V. Mikulec, M. Kuno, M. Bennati, D.A. Hall, R.G. Griffin, M.G. Bawendi, J. Am. Chem. Soc. 122 (2000) 2532-2540.

[11] L. Zu, D.J. Norris, T.A. Kennedy, S.C. Erwin, A.L. Efros, Nano Lett. 6 (2006) 15334

[12] O. Lehmann, K. Kompe, M. Haase, J. Am. Chem. Soc. 126 (2004) 14935-14942.

[13] J. Chen, M. Yao, X. Wang. J Nanopart Res 10 (2008) 163–171

[14] G. Murugadoss. J. Mater. Sci. Technol, 28 (7) (2012) 587–593.

[15] S.Wang, B.R.Jarrett, S.M.Kauzlarich, A.Y.Louie, J. Am. Chem. Soc., 129 (2007) 3848-3856.

[16] N.Pradhan, D. M. Battaglia, Y. Liu, X. Peng, Nano Lett., 7 (2007) 312-317.

[17] J.W. Stouwdam, R. A. J. Janssen. Advanced Materials 21 (2009) 2916–2920

[18] K.S. Rathore, Deepika, D. Patidar, N.S. Saxena, K. B. Sharma. Journal of Ovonic Research. 5 (2009) 175–185

[19] S.Salimian and S. Farjami Shayesteh. Acta Physica Polonica A. 118 (2010) 633-636

[20] N.Aangshuman, S. Sapra, S.S. Gupta, A. Prakash, A. Ghangrekar, N.Periasamy and D.D. Sarma. Bull. Mater. Sci. 31(2008) 561–568.

[21] S. Chandramohan, A. Kanjilal, S.N. Sarangi, S. Majumder, R. Sathyamoorthy, T. Som. J. Appl. Phys. 106 (2009) 063506-063511.

[22] B. Srinivasa Rao, V. R.Rajagopal, B. R. Kumar, T. Subba Rao, Int. J. Nanosci. 11 (2012) 1240006

[23] P. Hohenberg, W. Kohn, Phys.Rev. 136 (1964) B864-B871

[24] W. Kohn, L.J.Sham Phys.Rev. 140 (1965) A1133-A1138

[25] D. J. Singh, Plane Waves, Pseudopotentials and the LAPW Method (Kluwer Academic Publishers, Boston, 1994).

[26]P. Blaha, K. Schwarz, G. K. H.Madsen, D. Kvasnicka, and J. Luitz, WIEN2k, an Augmented Plane WaveRLocal Orbitals Program for Calculating Crystal Properties (Technische Universitat Wien, Austria, 2001).

[27] J. P. Perdew and Y. Wang, Phys. Rev. B, vol. 45 (1992) 13244-13249

[28] A.D. Becke, Phys. Rev. A, American Physical Society, 38 (1988) 3098-3100

[29] R. Seoudi, A.A. Shabaka , M. Kamal, E.M. Abdelrazek , W. Eisa . Physica E45 (2012) 47–55

[30] N.Soltani, E. Gharibshahi, E. Saion. Chalcogenide Letters. 9 (7) (2012), 321–328

[31] G.Xing Liang, P.Fan, Z.H. Zheng, J.T. Luo, D.P. Zhang, C.M. Chen, P.J. Cao. Applied Surface Science 273 (2013) 491–495

[32] Y.M. Yu, K.M. Kim, O. Byungsung , K.S. Lee, Y.D. Choi, and PY. Yu . J. Appl. Phys. 92 (2002) 1162

[33] F.D Murnaghan, Proc.Nat.Acad.Sci.USA 30 (1944) 244.

[34] A.Abbad, W.Benstaali, H.A. Bentounes, S.Bentata and A.Belaidi. Computational Materials Science 70 (2013) 19-23

[35] M.A. Khan, A. Kashyap, A. K. Solanki,T. Nautiyal, and S. Auluck, Phys. Rev. B 48 (1993) 16974-16978.

[36] B. Amin, I. Ahmad, M. Maqbool, N. Ikram, Y. Saeed, A. Ahmad, and S. Arif, Journal of Alloys and Compounds, 70 (2010) 874-880.

[37] F. Wooten, "Optical Properties of Solids," Academic Press, New York, 1972

[38] M. Fox, "Optical Properties of Solids," Oxford University Press, 2001.

[39] M. Thambidurai, N. Muthukumarasamy, S. Agilan, N. Sabari Arul, N. Murugan, R. Balasundaraprabhu, J. Mater. Sci. 46 (2011) 3200-3206

[40] B.Srinivasa Rao, B.Rajesh Kumar, V.Rajagopal Reddy, T.Subba Rao, G.Venkata Chalapathi. Chalcogenide Letters. 8 (2011) 39–44

FIGURES CAPTIONS

Fig.1: Total Density of States (a) $Cd_{16}S_{16}$ (b) $NiCd_{15}S_{16}$ (c) $Ni_2Cd_{14}S_{16}$. Fermi level is set to zero.

Fig.2: Partial Density of States (a) $Cd_{16}S_{16}$ (b) $NiCd_{15}S_{16}$ (c) $Ni_2Cd_{14}S_{16}$. Fermi level is set to zero.

Fig.3: Schematic representation of the supercells. (a) Conf. I (b) Conf. II (c) Conf. III

Fig.4: Total Density of States of $Ni_2Cd_{14}S_{16}$. (a) Conf. I (b) Conf. II (c) Conf. III Fermi level is set to zero.

Fig.5: Partial Density of States of $Ni_2Cd_{14}S_{16}$ (a) Conf. I (b) Conf. II (c) Conf. III Fermi level is set to zero.

Fig.6: Real and Imaginary part of the dielectric function of $Cd_{16}S_{16}$ and $Ni_2Cd_{14}S_{16}$ for the three configurations.

Fig.7: Absorption Coefficient of $Cd_{16}S_{16}$ and $Ni_2Cd_{14}S_{16}$ for the three configurations.

Fig.8: Reflectivity of $Cd_{16}S_{16}$ and $Ni_2Cd_{14}S_{16}$ for the three configurations.

TABLE

Table1: The Energy Difference ΔE between AFM and FM States (in meV), the scissor (in eV) used and the Ni-Ni Distances (in Å) for the three different configurations.

Configuration	$\Delta E = E_{AFM} - E_{FM}$ (meV)	coupling	d_{Ni-Ni} (Å)	ΔScissor (eV)
I	42	FM	4.07	2.2
II	35	FM	5.75	1.96
III	26	FM	8.14	1.86

TABLE 1

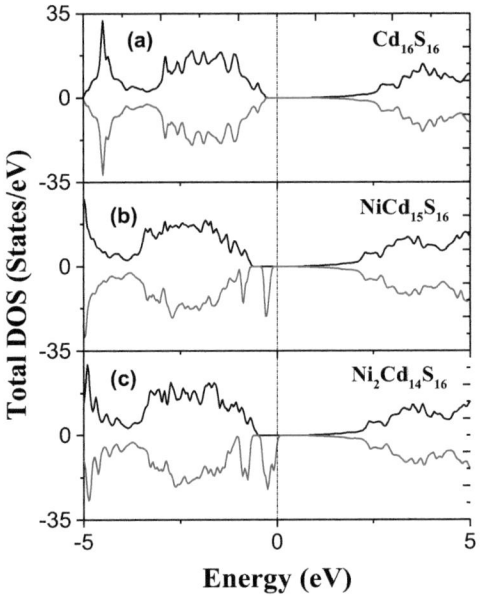

Fig. 1.

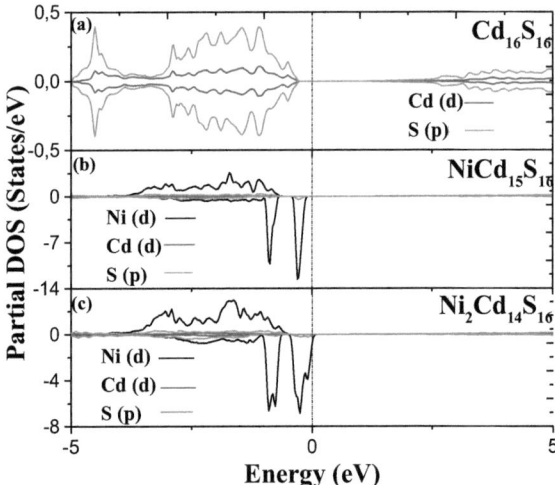

Fig. 2.

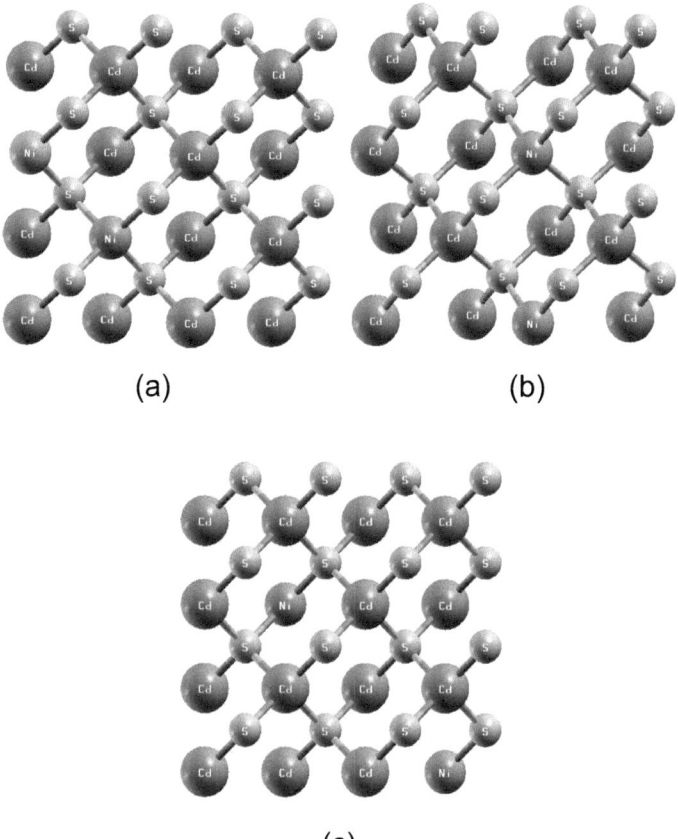

(a)   (b)

(c)

Fig. 3.

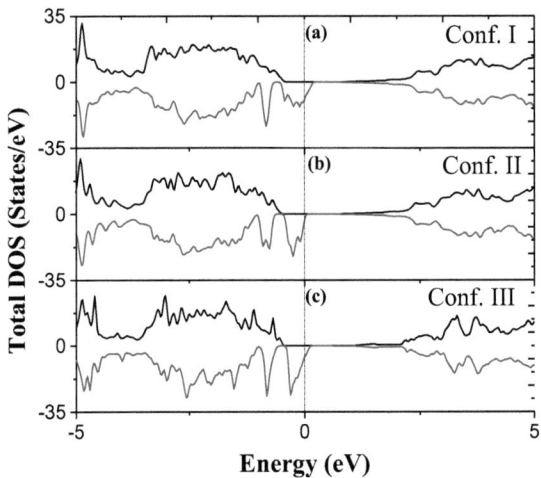

Fig. 04

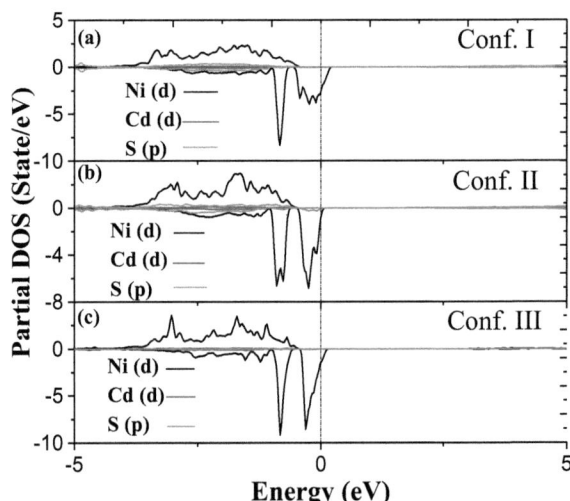

Fig. 5.

First-Principles Calculations of Magnetic, Electronic and Optical Properties of binary GaN and ternary CrGaN, CuGaN

A. Abbad*[,1], W. Benstaali[1], H.A. Bentounes[2,] S. Bentata[1] and A. Belaidi[3]

[1]Laboratory of Material Valorisation
Faculty of Sciences and Technology, BP227
Abdelhamid Ibn Badis University, Mostaganem (27000) Algeria
[2] Signals and Systems Laboratory (LSS)
Faculty of Sciences and Technology, BP227
Abdelhamid Ibn Badis University, Mostaganem (27000) Algeria
[3] Automatic and Systems Analysis Laboratory (LAAS), ENSET, Oran 31000, Algeria
*E-mail: am.ben@voila.fr

Keywords: density functional theory, Cr-doped GaN, Cu-doped GaN, ferromagnetism, half metallic, optical properties

Abstract

Using the full potential linear augmented plane wave (FPLAPW) method based on density functional theory (DFT), we have studied the magnetic, electronic and optical properties of Cr and Cu-doped GaN. The results show that GaN doped with non magnetic elements show ferromagnetism behaviour. The magnetic study shows that Cr doped GaN and Cu doped GaN are ferromagnetic but with different origin of magnetism for each compound. The electronic study indicates that Cr doped GaN is half metallic and n-type and that Cu doped GaN is also half metallic but P-type. Furthermore, we have computed optical properties of pure, Cr and Cu-doped GaN. By doping GaN, we found a pronounced peak occurring at low energies and an expansion of the curves for all the optical properties.

1. Introduction:

The search for magnetic materials with optimized optical, electronic, structural and magnetic properties has attracted much attention. Using the charge and spin degrees of freedom of electron in a single device are the main objective of the field of spintronics. The coupling between the charge carriers in a semiconductor and the electron spins of a ferromagnetic metal doped into the semiconductor can be utilized for many magneto/spin-electronic devices. The ultimate success of these materials for applications in "spintronics" relies on identifying the best candidates for this reason. The compound GaN has been extensively studied because of its promising applications in short wavelength opto-electronic devices [1]. GaN is a binary III-V direct band-gap semiconductor commonly used in bright light-emitting diodes since

17

the 1990s. Its wide <u>band gap</u> of 3.4 <u>eV</u> [2] affords it special properties for applications in <u>optoelectronic</u>, high-power and high-frequency devices. Among the various DMSs, transition metal doped GaN materials are particularly interesting and are regarded as prime candidates for spintronics applications. Since the theoretical prediction of room-temperature ferromagnetism in Mn-doped GaN [3, 4], wide research efforts have been devoted to the study of this system [5–14]. In many of these reports ferromagnetic behavior has been claimed to exist at or above room temperature. However, others have either reported ferromagnetism at only very low temperatures [15] or have provided evidence that the high TC ferromagnetism originates from clusters [16, 17]. The magnetic properties of Mn-doped GaN are found to depend strongly on the sample preparation conditions and diverse experimental groups have reported some contradictory results. Since the origin of ferromagnetism in Mn-doped GaN is difficult to be clearly identified, we look for the possibility of other 3d transition metal to dope GaN or we use other dopants which are intrinsically nonmagnetic but can be incorporated into semiconductors to form DMSs. Because such dopants are intrinsically nonmagnetic, their precipitates do not contribute to ferromagnetism. Cr is a neighbor to Mn in the Periodic Table, and several groups have recently reported above-room-temperature ferromagnetism for CrGaN [18, 19]. Also the possibility of Cu as a nonmagnetic dopant to fabricate GaN based DMS were examined [20, 21], but to the best of our knowledge, there has been no comparative study reported so far between these two elements doped- GaN, it should be noted also that the optical properties of transition metals doped GaN have not been extensively studied theoretically, this is why we report in this letter large investigation on optical properties of Cu and Cr-doped GaN after the study of their electronic and magnetic properties.

2. Calculation:

The electronic, magnetic and optical properties were studied by using the WIEN2K code [22, 23]. It is based on the full-potential linearized augmented plane wave method (FPLAPW) [24]. The electrons exchange-correlation energy is described in the local spin density approximation (LSDA) [25-26]. The muffin-tin (MT) radii of N, Cr, Cu and Ga were chosen to be 1.60, 1.90, 2.00 and 2.00 respectively. Basis functions were expanded as combinations of spherical harmonic functions inside non-overlapping spheres around the atomic sites (MT spheres) and in Fourier series in the interstitial region. We use 200 k points in the first Brillouin zone (74 special k points in the reduce wedge corresponding to a mesh of 7x7x3), this value of k point is obtained after an energy convergence test. The wave functions in the interstitial region were expanded in plane waves with a cutoff of $k_{max}=8/R_{MT}$ (where R_{MT} is the average radius of the MT spheres). The muffin-tin radius R_{MT} is based on two conditions: (i) no core charge leaks out of MT spheres and (ii) no overlapping is permitted between spheres. The wurtzite structure of GaN was used. The supercell employed contains 32 atoms (fig 1(a)), which correspond to a (2×2×2) supercell of GaN. The calculated equilibrium lattice parameters which minimize the total energy as a function of the cell volume are (in Å radii) a=3.1786 and c=5.1140 which are in good agreement with experimental and theoretical values [13, 27], these calculated values are obtained by fitting the total energy to Murnaghan's equation of state [28] for structural optimization. Because, the distribution of the TM into the structure play an important role in determining the electronic properties, we have first, for each doped structure, studied different configurations by changing the atomic position of the TM which were assumed to occupy Ga site and that corresponds to 6.25 at% concentration of TM, we have then chosen to use in our calculations the configuration that gave us the lowest minimum energy. The geometries of the supercells with and without TM doping were fully relaxed. Figure. 1 shows the supercell used in our calculations and for the doping process, the Ga atom in green is replaced by TM impurity. The optical properties were calculated using the Optic code implemented in Wien2k. Recognizing that the LSDA underestimates the band gap (see fig 2 later), we took into account in our optical calculations, for each studied structure (fig 1), an operator scissors which corresponds to the difference between the measured band-gap (3.40ev) and the calculated one.

3. Results and discussions

3.1 Electronic and magnetic properties:

The total density of states (DOS) for pure GaN is plotted in Fig. 2(a). From the DOS, we can see that the Fermi level is located in the gap region, confirming that GaN is a semiconductor. The DOS curves for spin-up and spin-down states are totally symmetric. The total and partial DOSs of Cr-doped GaN are shown in Figs. 2(a) and 3(b), respectively. Compared with pure GaN, Figs. 3(a) and 3(b) show remarkable changes of electronic structure in one substitutional Cr doped GaN. Some Cr correlative impurity levels are introduced near the bottom of conduction band, which modulates the conduction band minimum CBM and moves downward. The Fermi level (EF) is pinned in the conduction band, which means Cr-doped GaN is of n-type conductivity character. From the DOS and PDOS of Cr-doped GaN in Fig. 2(b), 3(a) and 3(b) respectively, we can predict behaviours of excess electrons introduced by doping because they might be deeply influenced by their energy level originating from the doped material. We note that the system is half-metallic and ferromagnetic since the Fermi energy passes through the DOS of spin down only. Meanwhile, there is a visible overlap between Cr-3d and N-2p states. In particular, the minority N-2p orbitals are more hybridized with Cr-3d orbitals than the majority states. We observe that the energy gap is reduced due to the presence of local strains and local electric fields generated by Cr atoms, suggesting that Cr-doping can enhance electronic conductivity. Concerning the n-type character of Cr-doped GaN mentioned above, we can conclude that the large PDOS values of Cr and its delocalized nature are considered to be the reasons of good conductivity of CrGaN and Cr 3d states contribute excess electrons as major carriers.

For an insight into the magnetism in Cr-doped GaN, our calculations show that $Ga_{15}CrN_{16}$ has a ferromagnetic ground state with a localized magnetic moment of 2.33 μ_B on the Cr atom. The neighboring N atom of Cr is polarized antiferromagnetically with a magnetic moment of 0.01 μ_B, which mainly comes from the N-2p orbital and there is no contribution from Ga 3d orbitals to the magnetic moment. So, we suggest that the double-exchange mechanism is responsible for ferromagnetism observed in Cr-doped GaN. In Table 1 we list the magnetic moments at each TM atom and its nearest-neighbours N and Ga atoms and interstitials.

For Cu-doped GaN the total and partial DOSs are shown in Figs. 2(c) and 3(c), respectively. We can see a half metallic behaviour because the majority spin is semiconducting and the minority spin is metallic. The Fermi level passes in the valence band which corresponds to a partial hole doping and the material is p-type. Our results are in good agreement with experimental ones reported by Wang [29]. The 2p state of the three connecting N atoms with Cu (Fig. 1) contributes significantly to the unoccupied states. In fact, each N atom has a magnetization around 0.19 μ_B

which is much larger than that of N in Cr-doped GaN as discussed above, elsewhere the (MM) of Cu is (0.65 μ_B) this indicate a strong hybridization between Cu and its three neighboring N atoms. This strong hybridization induces a finite magnetization on Cu atom as well as the neighboring N atoms, as shown in Fig. 3(c). Based on Zener's p-d hybridization mechanism for ferromagnetism in DMS [30, 31] , we imagine that it is the p-d hybridization mechanism that is responsible for ferromagnetism in Cu-doped GaN and therefore Cu should be a promising nonmagnetic dopant for GaN to fabricate GaN based DMS which promises to be free of magnetic precipitates.

As noted above (in calculation), the band gap of GaN is underestimated by the LSDA. The FPLAPW calculation gives E_g=2.10eV which is much lower than the measured one (3.40eV), so we have calculated the optical properties under the scissors operator. The scissors operator applied was 1.30 eV, which accounts for the difference between the experimental band gap and the calculated one of pure GaN.

3.2 Optical properties:
A number of phenomena can happen as the light propagates through the solids. They can be quantified by a number of parameters that determine their optical properties [32-35]. The most important optical property is the dielectric function $\varepsilon(\omega)$ which describes the polarization and absorption properties of the material. It is given by:

$$\varepsilon(\omega) = \varepsilon_1(\omega) + i\varepsilon_2(\omega)$$

Where $\varepsilon_1(\omega)$ and $\varepsilon_2(\omega)$ are the real and the imaginary parts of the dielectric function respectively.

The computed results for the real and imaginary parts of the dielectric function $\varepsilon(\omega)$ of pure GaN, Cr and Cu doped GaN are presented In Fig. 4. This figure shows the appearance of pronounced peaks with different heights at low energies at the band gap, in both real and imaginary parts, which are mainly due to the introduction of TM. Indeed, if we refer to the DOS on Figures 3 (a) and 3 (c), we find that the influence of Cr (3d) in Cr doped GaN is greater than that of Cu (3d) in Cu doped GaN, this is the same observation we see on Figure 4(a) and 4 (b) and this is why we can say that the two peaks situated in the infrared, are due to the free electrons and holes generated by the doping process and this effect is called "free carrier absorption". Also in Cu-doped GaN, there are three peaks in the infrared, because it is a p-type semiconductor. P-type semiconductors show an additional absorption mechanism in low energies due to intervalence band transitions [35]. These peaks which were not present in the pure structure of GaN demonstrate the interest of TM doping. In fact, these peaks show that there is a strong energy absorption which offers the possibility of using these doped compounds in the infrared for specific applications. We also remark that there is a shift of the doped curves to high energies.

The static dielectric constant ε_1 (0), which is the low energy limit of $\varepsilon_1(\omega)$ is robustly dependent on the bandgap of the compound. The inverse relation of ε_1 (0) with the band gap can be explained by Penn Model [36]. As ε_1 (0) augments, the band gap decreases and we find the lower value of band gap for Cr-doped GaN (see figure 2). Moreover, $\varepsilon_1(\omega)$ becomes zero at certain energy and then decreases to minimum values, with negative numbers, about 10.82eV, 11.62eV and 11.16eV for pure GaN, Cr, and Cu doped GaN respectively. The negative values of ε_1 (ω) show that the incident electromagnetic waves, in this energy region, are entirely reflected, consequently, the material shows metallic nature. When the energy increases after these critical points, ε_1 (ω) increases and becomes nearly zero at high energy limits (after 20ev). Furthermore, it is important to know, that for $Ga_{15}CrN_{16}$ and $Ga_{15}CuN_{16}$, the real part of the dielectric function is also negative in a very small region in the lower energies between 2.17eV and 2.29eV and between 2.82eV and 3.32eV respectively, these two regions which not exist in the pure structure must be taken into account for any eventual application because the behavior of the compound changes.

The imaginary part of the dielectric function ε_2 (ω) (Fig. 4) shows that the width of the absorption curve for $Ga_{15}CrN_{16}$ and $Ga_{15}CuN_{16}$ increases by a shift to lower and higher energies. We also note that the sharp peaks of pure GaN situated in the region between 7eV and 12.5eV are less apparent after TM doping.

The absorption coefficient for the pure and TM-doped GaN is represented in Fig 5. The absorption is mainly determined by the imaginary part of the dielectric function. The figure reveals several prominent peaks above absorption edge for pure GaN (E_g=3.33eV), which are due to excitonic effects.

These appeared peaks change to broad ones and are less pronounced for Cu and Cr-doped GaN,and are due to interband transitions of TM impurities from the topmost

VB to the lowest CB (Fig. 2). From the figure it is clear that the peak which corresponds to maximum absorption coefficient decreases after doping and is shifted towards high energies. Furthermore, we can see an important peak in the absorption coefficient centred at 2.16eV for $Ga_{15}CuN_{16}$ and 2.92eV for $Ga_{15}CrN_{16}$ which are generated by free carriers which are present in semiconductors at room temperature through the thermal excitation of electrons across the band gap[35] or by the presence of TM impurities. These two peaks explain the infrared absorption of maximum light at two different wave lengths. $Ga_{15}CuN_{16}$ and $Ga_{15}CrN_{16}$ have both significant absorption in large range of photon energies, and consequently, they are appropriates for device applications in the major parts of the spectrum.

Figure (6) show the reflectivity $R(\omega)$ versus energy of the three compounds. We see that the reflectivity is very close to 53 % for $Ga_{15}CrN_{16}$ and 40% for $Ga_{15}CuN_{16}$ in the infrared. In this region of low frequencies, pure GaN have low reflectivity (13%) and reaches its maximum value (60%) for 17.8eV, and then it drops sharply in the ultraviolet. The peaks at 3.00eV and 1.78eV for Cr and Cu doped GaN in the reflectivities curves are clearly related to the two peaks that appear at the infrared in the dielectric function. In pure GaN, different sharp peaks can be seen in the region from 6.5eV to 24.5eV which are due to excitonic effects, but after TM-doping they are less pronounced and we note that the curves are broadened to upper and lower energies. Moreover, the reflectivities are significantly enhanced in the energy range between 12 eV and 26 eV and arise from interband transitions, since both the real part of the dielectric function $\varepsilon_1(\omega)$ and the imaginary part $\varepsilon_2(\omega)$ are close to zero in this energy regime, (see Fig. 4). The local maximum values of reflectivity for Cr and Cu doped GaN at 3.00eV and 1.78eV are directly related to the negative values of $\varepsilon_1(\omega)$. A further attractive characteristic of the figure is the shift of the doped curves towards lower and higher energies. The frequency of optical peaks is proportionally related to the carrier density N of free electrons or holes generated by the doping process [37], via the next relation:

$$\omega_p^2 = \frac{Ne^2}{\varepsilon_0 m}$$

So, we can suggest that the carrier density N is the most important parameter which makes that we obtain a different shift of optical peaks for each TM impurity. The doped materials have a large reflectivity in a wide energy range than the undoped compound, in particular Cr-doped GaN which has the largest value of the width of reflectivity.

4. Conclusion:

We have studied the electronic, magnetic and optical properties of pure Cr, and Cu-doped GaN using the density functional theory within local spin density approximation for exchange and correlation potential. The magnetic study show that $Ga_{15}CrN_{16}$ has a ferromagnetic ground state with a localized magnetic moment on the Cr atom and we can possibly think that the double-exchange mechanism is one of the most important physical phenomenon responsible for ferromagnetism observed in Cr-doped GaN. The ferromagnetism exhibited makes Cr-doped GaN a promising candidate for applications in spintronics. However, we assume that the p-d hybridization mechanism is responsible for ferromagnetism in Cu-doped GaN. So, Cu should be a promising nonmagnetic dopant for GaN to fabricate GaN based DMS which promises to be free of magnetic precipitates. The electronic study show that the two doped compounds are half metallic and they are of different type. The study of optical properties (dielectric function, reflectivity and absorption coefficient) show that after doping, we have the emergence of significant peaks in the real and the imaginary parts of the dielectric function at low energies, which are originated from some facts discussed above, we also found that the bandwidth of the absorption coefficient increases, consequently we suggest that Cr and Cu doped GaN could be used for device applications in the major parts of the spectrum (infrared, visible and ultraviolet).

References:

[1] H. Morkoc, Nitride Semiconductors and Devices, Springer, Berlin, 1999.

[2] Streetman, G. Ben, S. Banerjee, Solid State electronic Devices (5th ed). New Jersey: Prentice Hall (2000) 524.

[3] H. Ohno, Science **281**, 951 (1998)

[4] T. Dietl, H. Ohno, F. Matsukura, J. Cibert, and D. Ferrant, Science **287**, 1019 (2000)

[5] K. Ando, Appl. Phys. Lett. **82**, 100 (2003)

[6] M. Luo, Z. Tang, Z. Q. Zhu, and J. H. Chu, J. Appl. Phys. **109**, 123720 (2011)

[7] S. Dhar, O. Brandt, A. Trampert, K. J. Friedland, Y. J. Sun, and K. H. Ploog, Phys. Rev. B **67**, 165205 (2003)

[8] M. Junaid, C.-L. Hsiao, J. Palisaitis, J. Jensen, P. O. A. Persson, L. Hultman, and J. Birch, Appl. Phys. Lett. **98**, 141915 (2011)

[9] S. Dhar, O. Brandt, A. Trampert, L. Däweritz, K. J. Friedland, K. H. Ploog, J. Keller, B. Beschoten, and G. Güntherodt, Appl. Phys. Lett. **82**, 2077 (2003)

[10] M. A. Załska-Kotur, F. Krzyzewski, and S. Krukowski, J. Appl. Phys. **109**, 023515 (2011)

[11] G. T. Thaler, M. E. Overberg, B. Gila, R. Frazier, C. R. Abernathy, S. J. Pearton, J. S. Lee, S. Y. Lee, Y. D. Park, Z. G. Khim, J. Kim, and F. Ren, Appl. Phys. Lett. **80**, 3964 (2002)

[12] P. P. Chen, H. Makino, J. J. Kim and T. Yao, J. Cryst. Growth **251**, 331 (2003)

[13] GuiQin Huang, and JiXia Wang, J. Appl. Phys. **111**, 043907 (2012)

[14] D. Ferrand, S. Marcet, W. Pacuski, E. Gheeraert, P. Kossacki, J.A. Gaj, J. Cibert, C. Deparis, H. Mariette, C. Morhain, J. Superconductivity **18**, 15 (2005)

[15] K. Edmonds, S.V. Novikov, M. Sawicki, R. Campion, C.R. Staddon, A.D. Giddings, L.X. Zhao, K.Y. Wang, T. Dietl, C.T. Foxon, B.L. Gallagher, Appl.Phys. Lett. **86**, 152114 (2005)

[16] T. Graf, M. Gjukic, M. Hermann, M.S. Brandt, M. Stutzmann, L. Gorgens, J.B. Philipp, O. Ambacher, J. Appl. Phys. **93**, 9697 (2003)

[17] M. Zajac, J. Gosk, E. Grzanka, M. Kaminska, A. Twardowski, B. Strojek, T. Szyszko, S. Podsiadlo, J. Appl. Phys. **93**, 4715 (2003)

[18] JungHwan Chun and Dong Eon Kim. Phys. Status Solidi A. **208**, 691-694 (2011)

[19] K. Sato and H. Katayama-Yoshida, Jpn. J. Appl. Phys. Part 2 **40**, L485 (2001)

[20] A. L. Rosa and R. Ahuja Appl. Phys. Lett. **91**, 232109 (2007)

[21] H. J. Xiang and Su-Huai Wei Nano Lett., **8** , 1825–1829 (2008)

[22] K. Schwarz and P. Blaha, Computational Materials Science. **28**, 259 (2003)

[23] P. Blaha, K. Schwarz, G. K. H. Madsen, D. Kvasnicka, J. Luitz, WIEN2K-An Augmented plane wave & Local Orbital Program for Calculating Crystal Properties (Techn. Universitat Wien, Austria, 2001).

[24] O. K. Andersen, Phys . Rev. B **12**, 3060 (1975).

[25] A.D. Becke, Phys. Rev. A, American Physical Society, vol. 38 n° 6 (1988) 3098-3100

[26] Pushpendra Kumar and Kedar Singh, Journal of optoelectronic and Biomedical Materials 1 (2009) 59-69.

[27] C. G. Liang and J. Zhang, Chin. J. Semicond. **20**, 89 (1999)

[28] F.D Murnaghan, Proc.Nat.Acad.Sci.USA 30 (1944) 244.

[29] Wang Peng-Wei *et al, Chinese Phys. Lett.* **25** 3040 (2008)

[30] H. Akai, Phys. Rev. Lett. 81, 3002 (1998).

[31] K. Sato, P. H. Dederichs, H. Katayama-Yoshida, and J. Kudrnovsky, J.Phys. Condens. Matter 16, S5491 (2004).

[32] M. A. Khan, A. Kashyap, A. K. Solanki,T. Nautiyal, and S. Auluck, Phys. Rev. B **48**, 16974 (1993)

[33] B. Amin, I. Ahmad, M. Maqbool, N. Ikram, Y. Saeed, A. Ahmad, and S. Arif, Journal of Alloys and Compounds, **70**, 874 (2010)

[34] F. Wooten, "Optical Properties of Solids," Academic Press, New York, 1972
C. Ambrosch-Draxl and J. O. Sofo, Comput. Phys. Commun. **175**,1 (2006)

[35] M. Fox, "Optical Properties of Solids," Oxford University Press, 2001

[36] D. Penn, Phys. Rev. 128 (1962) 2093

[37] N.W. Ashcroft and N.D. Mermin, Solid State Physics, (Saunders College, Philadelphia, 1976).

TABLE

Table1: Calculated Magnetic Moments (in Bohr Magneton µB) for Several Sites Of Each Structure

FIGURES CAPTIONS

Fig.1: Schematic representation of the supercell. The green sphere is the defect site.

Fig.2: Total Density of States (a) $Ga_{16}N_{16}$ (b) $Ga_{15}CrN_{16}$ (c) $Ga_{15}CuN_{16}$.

Fig.3: Partial Density of States (a) Cr(3d) and N(2p) for $Ga_{15}CrN_{16}$, (b) Ga(3d) and N(2p) for $Ga_{15}CrN_{16}$, (c) Cu(3d) and N(2p) for $Ga_{15}CuN_{16}$

Fig.4: Real and Imaginary part of the dielectric function of $Ga_{16}N_{16}$, $Ga_{15}CrN_{16}$ and $Ga_{15}CuN_{16}$

Fig.5: Absorption coefficient of $Ga_{16}N_{16}$, $Ga_{15}CrN_{16}$ and $Ga_{15}CuN_{16}$

Fig.6: Reflectivity of $Ga_{16}N_{16}$, $Ga_{15}CrN_{16}$ and $Ga_{15}CuN_{16}$

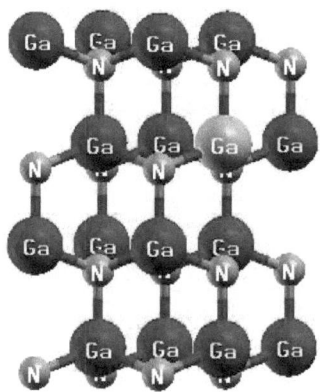

Fig.1

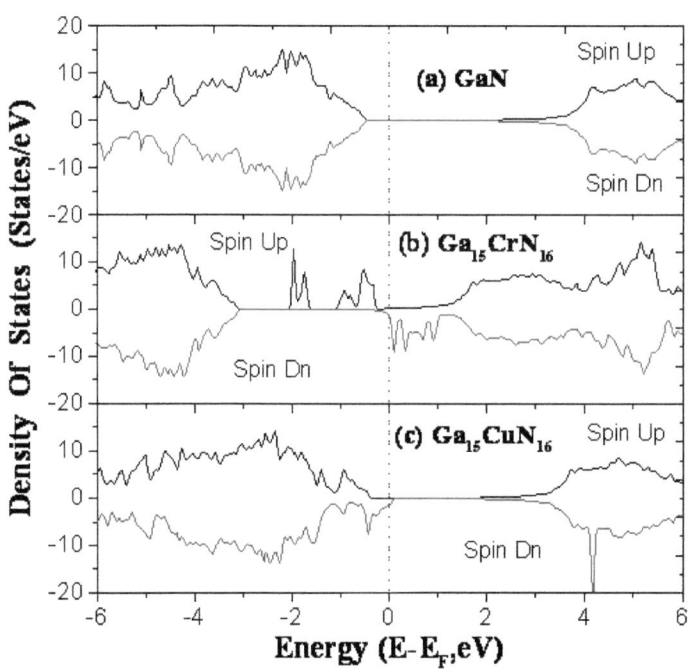

Fig.2

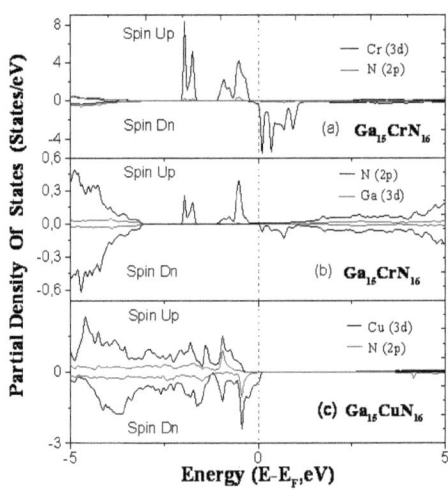

Fig.3

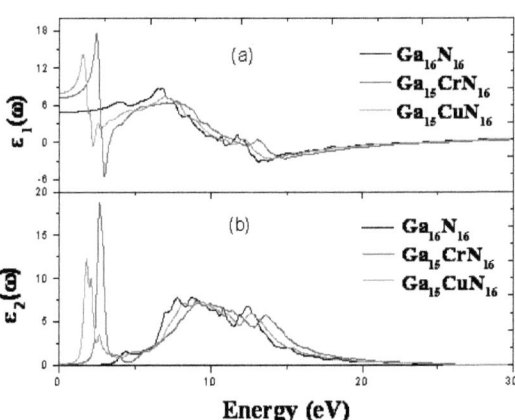

Fig.4

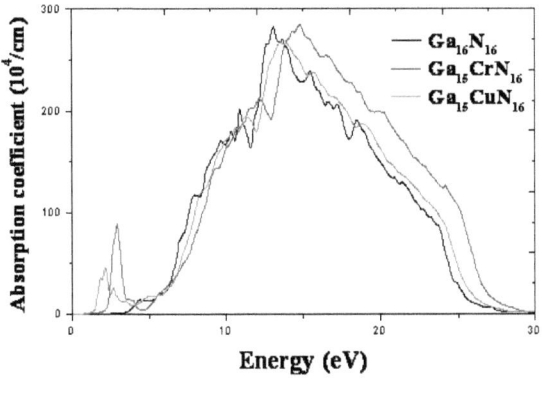

Fig.5

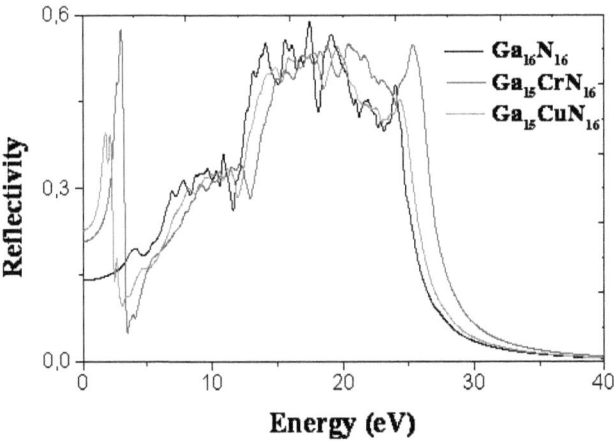

Fig.6

Site	Ga$_{15}$CrN$_{16}$	Ga$_{15}$CuN$_{16}$
Ga	0.001	0.004
N	-0.001	0.19
Cr	2.33	-
Cu	-	0.65
μ_{total}	3	1.90
$\mu_{interstitiel}$	0.58	0.26

Table 1

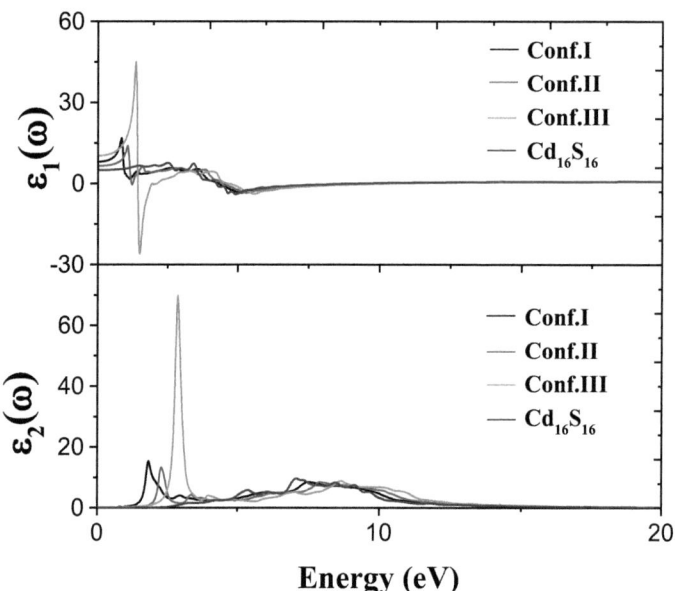

Fig.06

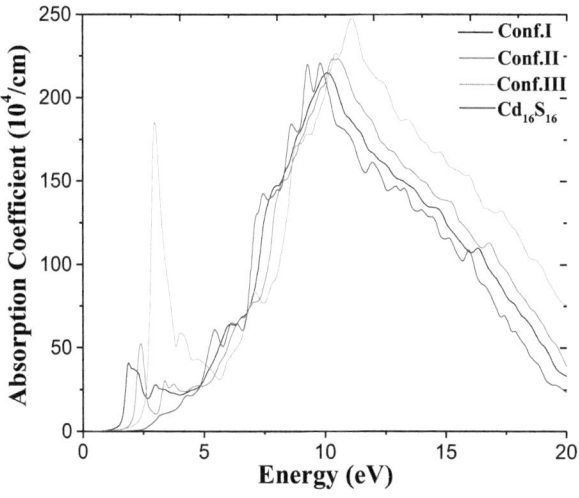

Fig. 07

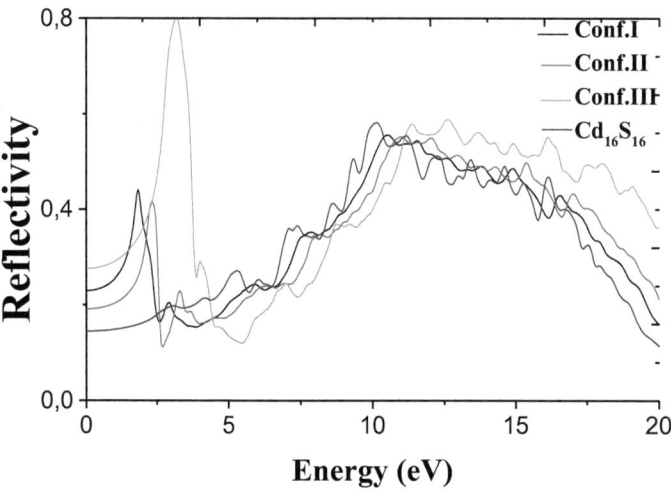

Fig. 08

Optoelectronic Properties of Transition Metals doped Cubic Cadmium Sulphide

W.Benstaali[1], S. Bentata[1], A. Abbad [2] , H.A.Bentounes[2] , and T. Lantri[1]

[1]Laboratory of Technology and of Solids Properties
Faculty of Sciences and Technology, BP227
Abdelhamid Ibn Badis University, Mostaganem (27000) Algeria
[2] Signals and Systems Laboratory (LSS)
Faculty of Sciences and Technology, BP227
Abdelhamid Ibn Badis University, Mostaganem (27000) Algeria

*E-mail:ben_wissam@yahoo.fr

Abstract

First principle calculations have been performed to obtain detailed electronic structure, and optical properties of Fe, Pd and Mn doped CdS. The results show that FeCdS and PdCdS show spin polarization phenomenon since they *have a metallic behavior for spin down and an insulating character for spin up. This half*-metallic character *makes these compounds potential candidates for applications in spintronics.* Moreover, optical properties show *the appearance of pronounced peak in PdCdS and FeCdS at low energies. Consequently, these peaks prove that there is a strong energy absorption which offers the possibility of using these doped compounds for specific applications.*

- **Keywords:** Total Density of states; Magnetic Moments; Optical Properties; First-principles calculations, Drude Model.

1. Introduction:

In recent years, the II–VI and III–V based semiconductors have attracted a great deal interest due to their potential use as photoconductors and in fabricating optical detectors, light sources and optoelectronic devices [1-10]. Their structural, electronic and optical properties have been widely studied in both theoretical and experimental works. One of the motives to consider their properties is the need to tune and control the energy band gap of semiconductors, which affords the base for reaching optimal range over which optoelectronic devices operate. Cadmium Sulphide, CdS belongs to the II–VI compound family with direct band gap energy of

2.42 eV [11]. CdS has been widely used as an n-type window material in thin-film solar cell devices. In particular, several recent applications using CdS sheets have been proven including high quality field effect transistors, optical switches, and light emitting diodes [12-17].

Diluted magnetic semiconductors (DMS) are semi conducting materials in which a fraction of the host cations can be substitutionally replaced by magnetic ions or appropriate rare earths. To discover diluted magnetic semiconductor (DMS) is a challenge in the field of spintronics. Compared to conventional semiconductors, spintronic devices join both the electronic and spin degrees of freedom in one material. Much of the interest on DMS materials is due to its potential application in "spintronics" devices, which exploit spin in magnetic materials along with charge of electrons in semiconductors [18]. After the prediction of ferromagnetism at room temperature (RT) by doping transition metal (TM) in wide band-gap II–VI and III–V compounds by Dietl et al. [19], numerous efforts have been made to search for RT magnetism in CdS by doping with TM elements [20-28]. Therefore, CdS DMSs are very promising materials for spintronic devices operating at high temperature. However, the origin of the magnetism of CdS based DMSs is still under active debate.

In the recent years, much effort has been invested in exploring electronic and optical properties of some transition metals doped CdS. However, no comparative study on TM-doped CdS is available to our knowledge. Therefore, it is of interest to find out whether Pd, Fe or Mn when incorporated in CdS, give best electronic and best optical properties than pure CdS, to get it the nature of doping (n-type or p-type) and to see the effect on magnetic properties.

2. Calculation:

All calculations are performed with the full potential linearized augmented plane wave (FPLAPW) method [29], embedded in wien2k code [30]. The generalized gradient approximation (GGA) [31] is used to treat the exchange correlation potentials. The parameter $R_{min}K_{max}$ (smallest muffin tin radius multiplied by the maximum k value in the expansion of plane waves in the basis set), which controls the size of the basis set in our calculation, is chosen as 7.5. Following the

Monkhorst–Pack scheme [32], a 4 × 9 × 9 k-point mesh is adopted. The calculated equilibrium lattice parameter which minimizes the total energy as a function of the cell volume was found (a=5.7384 Å). This value of (a) is in good agreement with experimental and theoretical ones [33, 34]. It was obtained by fitting the total energy to Murnaghan's equation of state [35] for structural optimization. In this study, one atom Cd is replaced by one TM as we can see on Fig.1 [36].

For computing optical properties, we used two methods. The first one for the insulating orientation of the spin and the second one for the metal orientation of the spin where we used the Drude model [37]. For this last calculation, the intraband contribution is taking into consideration, so we need to know the new values of plasma frequencies which can be obtained by taking a value of switch equal to 6 in the case.injoint file. Then, we recalculate the optical properties by taking a value of switch equal to 4 in the case.injoint file with putting the new values calculated of plasma frequencies in the case.inkram file (ω_{pxx} = 0.1602 and ω_{pzz} = 0.5533 for FeCdS and ω_{pxx} = 0.9246 and ω_{pzz} = 1.3113 for PdCdS), we must also take into account the value of scissor which corresponds to a metal.

3. Results and discussions

3. 1. Electronic Properties

The spin-polarized total density of states (DOS) of compounds calculated within GGA at the obtained equilibrium lattice constants are shown in Fig.2. The DOS of doped structures are given for spin up and spin down. For pure CdS, the Fermi level is located in the gap region, confirming that CdS is a semiconductor. The calculated results show that pure CdS has a direct band gap of about 1.03 eV, which is smaller than the experimental value (2.5 eV), because the GGA underestimates the band-gap. This is essentially due to the fact that it has simple form that is not sufficiently flexible to reproduce accurately exchange-correlation energy. From Fig.2, we can see a difference in the DOS of doped materials, depending on the orientation of the spin. The TM impurity introduces some states around the Fermi level, which change the electronic properties of semiconductor CdS. For FeCdS, some iron correlative impurity levels are introduced near the bottom of conduction band (CB), which modulates the conduction band minimum CBM and moves downward. The Fermi level (EF) is pinned in the conduction band, which means that Fe-doped CdS is of n-type conductivity character. However for PdCdS, the Fermi level passes in the valence band (VB) which corresponds to a partial hole doping and the materials are p-type. The FeCdS and PdCdS show spin polarization phenomenon because they have a metallic behavior for spin down, and an insulating character for spin up. This half-metallic character makes probably these compounds promising candidates for applications in spintronics. For MnCdS, we can see from Fig.2, that the Fermi level is situated through the band gap, this means that the compound keeps its semi conducting behaviour, we can also observe a reduction in the band gap compared to pure CdS which is due to the appearance of localized states for the spin down of the conduction band (CB), which move toward the low energies, and for the spin up of the valence band (VB) which move slightly toward the high energies. Therefore, MnCdS is an intrinsic insulator and it is unsuitable for spintronics except in the case where another type of impurity is added.

To intensely understand the electronic properties of the materials of interest, we have computed the partial density of states. Figs.3 presents the partial DOS of MnCdS, FeCdS, and PdCdS respectively. We note that for the FeCdS there is a visible overlap between TM-3d and S-3p states and there is also for PdCdS, a perceptible overlap between TM-4d and S-3p states. In particular, the spin down for S-3p orbitals are more hybridized with TM-3d and TM-4d orbitals respectively than the spin up states. This hybridization is more pronounced in the case of PdCdS and the peaks are less delocalized. We observe that the energy gap is reduced due to the presence of local strains and local electric fields generated by TM atoms, suggesting that TM-doping can enhance electronic conductivity.

For a further study of electronic properties, we calculated band structures of FeCdS and PdCdS (Fig. 4 and Fig. 5 respectively). We observe, for the insulating orientation of the two compounds, a direct band gap since the maximum of the valence band and the minimum of the conduction band are on the same line passing through the point of high symmetry Γ. The fact that we have a direct band gap, this means that these compounds should have attractive optical properties for the reason that the optical transitions are direct.

To confirm the half-metallic behavior of FeCdS and PdCdS, we have performed other calculations using GGA+U where U-Hubbard is the on-site Coulomb interaction which treats the localized d electron states in the iron and palladium atoms. The introduction of U-Hubbard in present calculations gives more accurate results because of the correction which is given to the band gap. Fig.6 (a) shows, for U=6eV, a shift of spin down states of the VB of FeCdS to CB and a breakup of the localized bands which were located around Fermi level when we used GGA approach, so this leads also to a metallic behavior. For spin up, we observe an insulating behavior with a band gap which is close to the experimental value. This means that FeCdS keeps the half-metallic character as for the GGA. On Fig. 6 (b), we observe that the spin down states of VB move to CB while for spin up, we have also an insulating character. Consequently the half-metallicity nature of the component is maintained.

For an insight into the magnetism in TM-doped CdS, our calculations using GGA show that FeCdS, MnCdS and PdCdS have a ferromagnetic ground state which is mainly due to the magnetic moment of TM. The remaining values of magnetic moments are listed in table 1, where we can see that Cd and S atoms also contribute to magnetic moment but with a very small part due to hybridization, we can also confirm the half metallicity of FeCdS and PdCdS because their total magnetic moment is an integer value (4µB and 2 µB respectively) .

3. 2. Optical Properties

A number of optical parameters can be obtained by Wien2k such as: the dielectric function, the refractive index, the reflectivity and the absorption coefficient [38-39].

In Fig.7, we present the computed results for the real part of the dielectric function of pure CdS, Mn, Fe and Pd doped CdS, for the two directions of spin. In Fig. 7(a) (spin up), for pure CdS, we can see a less peak which occurs at 3.77 eV and that should be mainly caused by the transition between S-3p orbital in higher valence band and Cd-5s in the lowest conduction band. We can also observe on Fig. 7(a) the appearance of a pronounced peak in doped material PdCdS at low energies, which is mainly due to the introduction of transition metal impurities. In fact, this peak proves that there is a strong energy absorption which offers the possibility of using these doped compounds for specific applications. We also note that for each TM, the peaks have different height values and are situated in different regions.

When we compare PdCdS and pure CdS on Fig.7 (a), we can say that the peak which was shown in pure CdS is found again in PdCdS, but it is offset to the lower energy side (red shift) and is wider than it was in pure CdS. These phenomena can be explained by the density of states, Pd atom provides a large number of conduction holes near the Fermi level, and so it changes the electronic interband transitions, and has an impact on the dielectric function and optical properties. The inverse occurs for FeCdS (blue shift), because the material is n type. In the case of MnCdS, the peak is situated in the same energy value as pure CdS, because there is no charges (electrons or holes), near Fermi level. Furthermore, $\varepsilon_1(\omega)$ becomes zero at certain energy and then decreases to minimum values, with negative numbers. The negative values of $\varepsilon_1(\omega)$ show that the incident electromagnetic waves, in this energy region, are entirely reflected, consequently, the material shows metallic nature [39].

On Fig. 7(b) (spin down), we treat PdCdS and FeCdS sush as metals, so we take into account the intraband transition by calculating the new values of plasma frequencies, the values of Drude correction corresponding to plasma frequencies and the scissor corresponding to the metal. An important peak with negative value for PdCdS at zero energy can be seen. After that, $\varepsilon_1(\omega)$ increase and become positive for certain energies. The peaks noted from the curves and for the two compounds PdCdS and FeCdS are due to a competition between interband and intraband transitions.

The imaginary part of the dielectric function $\varepsilon_2(\omega)$ (Fig. 8 (a)), shows that the width of the absorption curve for MnCdS, FeCdS and PdCdS, increases by a shift to higher energies. We also note that the sharp peaks of pure CdS situated in the region between 5eV and 10eV are less apparent after TM doping. The first critical points occur in the visible at 1.86 eV, 1.97 eV, 1.48 eV and 1 eV for MnCdS, pure CdS, FeCdS and PdCdS, respectively. These energies, represent, transition threshold between the maximum valance band and the minimum conduction band. Moreover, on Fig. 8 (b), we observe a different behaviour for FeCdS and PdCdS since we treat them as metals. At E=0 eV, two important peaks can be noted at 9.59 eV for FeCdS and at 19.53 eV for PdCdS.

The absorption coefficient for pure and TM-doped CdS is represented on Fig. 9 (a) for spin up and on Fig 9 (b) for spin down. We observe an appearance of different peaks which are directly linked to dielectric function and are essentially due to TM and inter band transition. From Fig. 9 (a), it is clear that absorption coefficient is blue-shifted after TM doping for FeCdS and PdCdS which have both, significant absorption coefficient in large range of photon energies, and consequently, they are appropriates for device applications in the major part of the spectrum. In the contrary, the absorption coefficient for these two compounds is red-shifted in spin down (Fig. 9 (b)). For spin up, the maximum absorptions are: $140*10^4*cm^{-1}$, $141*10^4*cm^{-1}$,

$138*10^{4*}cm^{-1}$, $112*10^{4*}cm^{-1}$, for pure CdS, FeCdS, PdCdS and MnCdS. These values are clearly reduced when intraband effects are considered, so we have $111*10^{4*}cm^{-1}$ for FeCdS and $104*10^{4*}cm^{-1}$ for PdCdS.

On Fig.10, we plot the different reflectivities $R(\omega)$ versus energy for the four compounds. The peaks in the reflectivities curves (spin up) are clearly related to peaks in the dielectric function. In pure CdS, different sharp peaks can be seen in the region from 5 eV to 17 eV but after TM-doping, they become less pronounced and we note that the curves are slightly broadened to upper and lower energies. Besides, the reflectivities for the doped compounds are significantly enhanced in the energy range between 8 eV and 20 eV as a result of inter-band transitions, since the absorption in dielectric function becomes smaller in the same area energy (see Fig. 9 (a)).On Fig. 10 (b) and at E=0eV, the reflectivity reach it maximum values; 36% for FeCdS and 85 % for PdCdS. A further attractive characteristic of the figure is the blue shift of the curves for FeCdS and PdCdS in the spin up. The doped materials have a larger reflectivity energy range than the undoped compound.

4. Conclusion:

In summary, we used the density functional theory in the generalized gradient approximation (GGA) in order to study the electronic and optical properties of some transition metals doped Cadmium sulphide. We have studied three different compounds in where one atom Mn, Fe, or Pd replaces one Cd atom in the supercell. The 3d electrons of Fe and 4d electrons of Pd make pure CdS change into half metallic semiconductor, narrow the band gap and show spin polarization, we have also performed GGA+U calculations which confirms the half-metallic behaviour of FeCdS and PdCdS, so these compounds can be promising materials for applications in spintronics. Additionally, Fe-doped CdS is of n-type conductivity character, and PdCdS is p-type. The 3d electrons of Mn, lead to the appearance of localized states in the minority spin of the conduction band (CB), which move toward the low energies, and in the majority spin of the valence band (VB) which move slightly toward the high energies, but show no spin polarization. For that reason, MnCdS is an intrinsic insulator and it is unsuitable for spintronics only if another type of impurity is added.

The study of optical properties (dielectric function, reflectivity and absorption coefficient) shows that after doping, we have the emergence of significant peaks in the real and the imaginary parts of the dielectric function. Theses peaks are originated from interband transition for spin up and intraband transitions for spin down for the two compounds FeCdS and PdCdS, we also found that the bandwidth of the absorption coefficient increases, consequently Fe and Pd doped CdS are appropriates for device applications in the major parts of the spectrum (infrared, visible and ultraviolet).

Acknowledgment

The authors thank Pr. B.Bouhafs and Pr B. Abbar for their helpful discussions.

References:

[1] J. Britt, C. Ferekides, Applied Physics Letters, 62 (1993) 2851.

[2] M. Savelli, J. Bougnot, Applied Physics, 31 (1979) 213.

[3] J. Kaur, D.K. Paudya, K.L. Chopra, Journal of the Electrochemical Society, 127 (1980) 943.

[4] P. Li, C. W. Zhang, J. Lian , M. J. Ren , P. J. Wang , X. H. Yu, S. Gao, Optics Communications, 295 (2013) 45.

[5] J. Jie, W. Zhang, I. Bello, C.S. Lee, S.T. Lee, Nano Today, 5 (2010) 313.

[6] J. Phillips, W. Bowen, E. Cagin, W. Wang, Comprehensive Semiconductor Science and Technology, 6 (2011) 101.

[7] Y. Al-Douri, Procedia Engineering, 53 (2013) 400.

[8] J. Fan, L. Ouyang, X. Liu, D. Ding, J K. Furdyna, D. J.Smith, Y.H. Zhang, Journal of Crystal Growth, 323 (2011) 127.

[9] C. Tablero, Computational Materials Science, 37 (2006) 483.

[10] M. Shkir, G. Bhagavannarayana, M.A. Wahab, K.K. Maurya, International Journal for Light and Electron Optics, 124 (2013) 1995.

[11] J. Cao, J. Sun, J. Hong, H. Li, H. Chen, M. Wang, Advanced Materials, 16 (2004) 84.

[12] X.F. Duan, C.M. Niu, V. Sahi, J. Chen, J.W. Parce, S. Empedocles, J. L. Goldman, Nature 425 (2003) 274.

[13] T. Gao, Q.H. Li, T.H. Wang, Applied Physics Letters, 86 (2005) 173105.

[14] R.M. Ma, L. Dai, H.B. Huo, W.Q. Yang, G.G. Qin, P.H. Tan, C.H. Huang, J. Zheng, Applied Physics Letters, 89 (2006) 203120.

[15] R.M. Ma, L. Dai, G.G. Qin, Nano Letters, 7 (2007) 868.

[16] Y. Jiang, J.Z. Wen, S.J. Jian, M.M. Xiang, X. Fan, S.T. Lee, Advanced Functional Materials, 17 (2007) 1795.

[17] J. S. Jie, W. J. Zhang, Y. Jiang, X. M. Meng, Y. Q. Li, S. T. Lee, Nano Letters, 6 (2006) 1887.

[18] A. Abbad, H. A. Bentounes, W. Benstaali, A. Belaidi, Journal of Magnetism and Magnetic Materials, 326 (2013) 28.

[19] T. Dietl, H. Ohno, F. Matsukura, J. Cibert, D. Ferrant, Science, 287 (2000) 1019.

[20] V. Ladizhansky, V. Lyahovitskaya, S. Vega, Physical Review B, 60 (1999) 8097.

[21] N. S. Norberg, D. R. Gamelin, Journal of Applied Physics, 99 (2006) 08M104.

[23] S. Aksu, E. Bacaksiz, M. Parlak, S. Yılmaz, I. Polat, M. Altunbas, M. Turksoy, R. Topkaya, K. Ozdogan, Material Chemistry and Physics, 130 (2011) 340.

[24] R. Sathyamoorthy, P. Sudhagar, A. Balerna, C. Balasubramanian, S. Bellucci, A.I. Popov, K. Asokan, Journal of Alloys and Compounds, 493 (2010) 240.

[25] T. Hu, M. Zhang, S. Wang, Q. Shi, G. Cui, S. Sun, Cryst. Eng. Comm.13 (2011) 5646.

[26] P. Srivastava, P. Kumar, K. Singh, Journal of Nanopartical Research, 13 (2011) 5077.

[27] K. A. Bogle, S. Ghosh, S.D. Dhole, V. N. Bhoraskar, L. Fu, M. Chi, N.D. Browning, D. Kundaliya, G. P. Das, S.B. Ogale, Chemistry of Materials, 20 (2008) 440.

[28] J.M.D. Coey, Solid State Science, 7 (2005) 660.

[29] D. J. Singh, Plane Waves, Pseudopotentials and the LAPW Method (Kluwer Academic Publishers, Boston, 1994).

[30] P. Blaha, K. Schwarz, G. K. H.Madsen, D. Kvasnicka, and J. Luitz, WIEN2k, an Augmented Plane WaveRLocal Orbitals Program for Calculating Crystal Properties (Technische Universitat Wien, Austria, 2001).

[31] J.P. Perdew, K. Burke, M. Ernzerhof, Phys. Rev. Lett. 77 (1996) 3865.

[32] H.J. Monkhorst, J. Pack, Phys. Rev. B 13 (1976) 5188.

[33] Y.M. Yu, K.M. Kim, O. Byungsung, K.S. Lee, Y.D. Choi, P.Y. Yu . Journal of Applied Physics, **92** (2002) 1162.

[34] M. A. Khan, A. Kashyap, A. K. Solanki,T. Nautiyal, and S. Auluck, Phys. Rev. B 48 (1993) 16974.

[35] F. D. Murnaghan, Proceedings of the *National Academy* of Sciences.USA, 30 (1944) 244.

[36] A. Kokalj, Comp. Mater. Sci., Vol. 28, p. 155, 2003.

[37] M. Fox, "Optical Properties of Solids," Oxford University Press, 2001

[38] B. Amin, I. Ahmad, M. Maqbool, N. Ikram, Y. Saeed, A. Ahmad, and S. Arif, Journal of Alloys and Compounds, 70 (2010) 874.

[39] F. Wooten, "Optical Properties of Solids," Academic Press, New York, 1972

FIGURES CAPTIONS

Fig.1: Structure of pure CdS and FeCdS. [(a) , (b)]

Fig.2: Total Density of States of CdS, MnCdS, FeCdS, and MnCdS. Fermi level is set to zero.

Fig.3: Partial Density of States of Density of States of MnCdS, FeCdS, and PdCdS. Fermi level is set to zero.

Fig.4: Band structure of FeCdS. (a): Spin Up, (b): Spin Down

Fig.5: Band structure of PdCdS. (a): Spin Up, (b): Spin Down

Fig.6: Total Density of States using GGA+U. (a): FeCdS, (b): PdCdS.

Fig.7: Real part of the dielectric function of CdS, MnCdS, FeCdS and PdCdS. (a) Spin Up (b) Spin Down.

Fig.8: Imaginary part of the dielectric function of CdS, MnCdS, FeCdS and PdCdS. (a) Spin Up (b) Spin Down.

Fig.9: Absorption Coefficient of CdS, MnCdS, FeCdS and PdCdS. (a) Spin Up (b) Spin Down.

Fig.10: Reflectivity of CdS, MnCdS, FeCdS and PdCdS. (a) Spin Up (b) Spin Down.

TABLE

Table1: Total and local magnetic moment (in μ_B), for each material studied

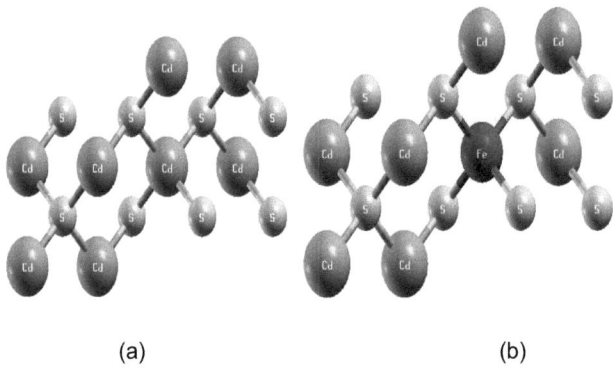

(a)   (b)

Fig. 1

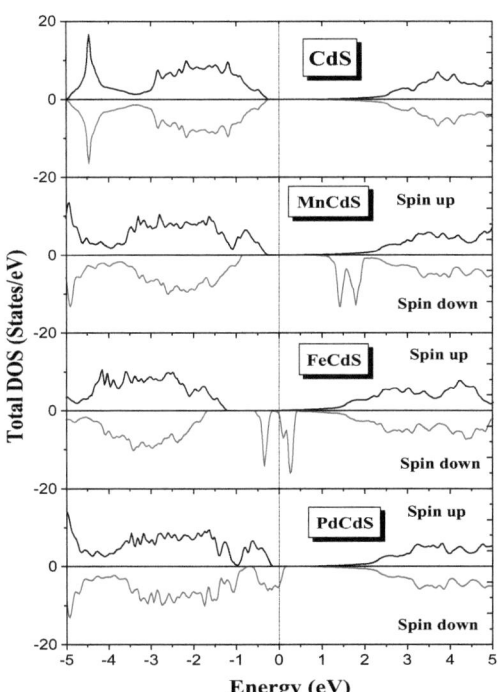

Fig.2

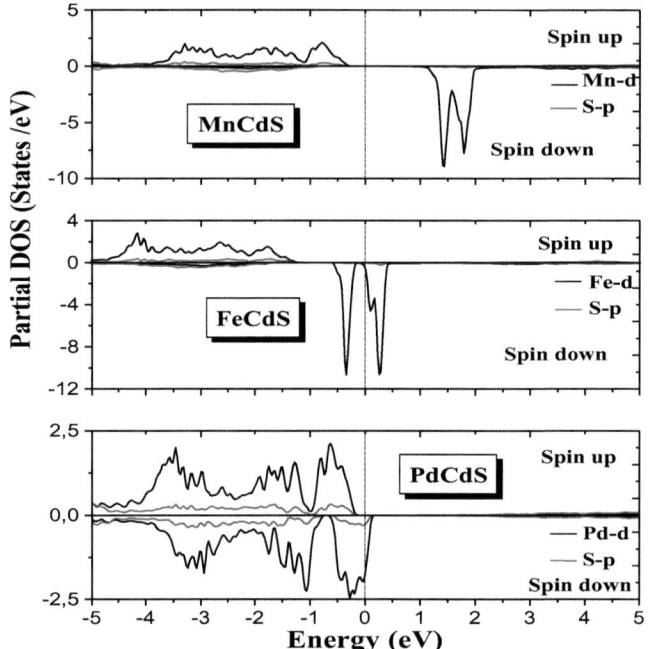

Fig. 3

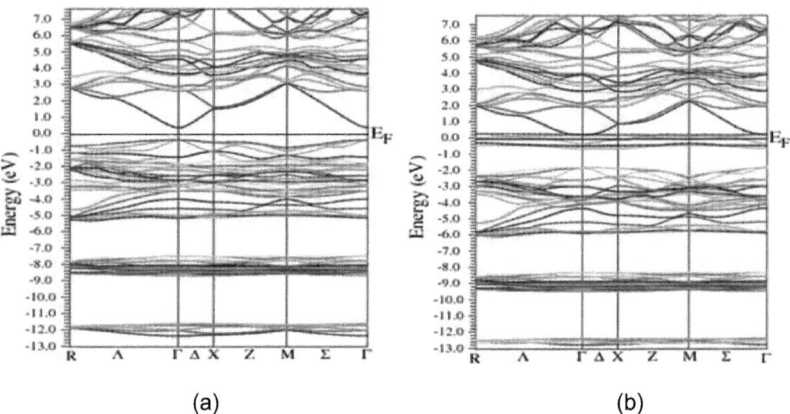

(a)   (b)

Fig. 4

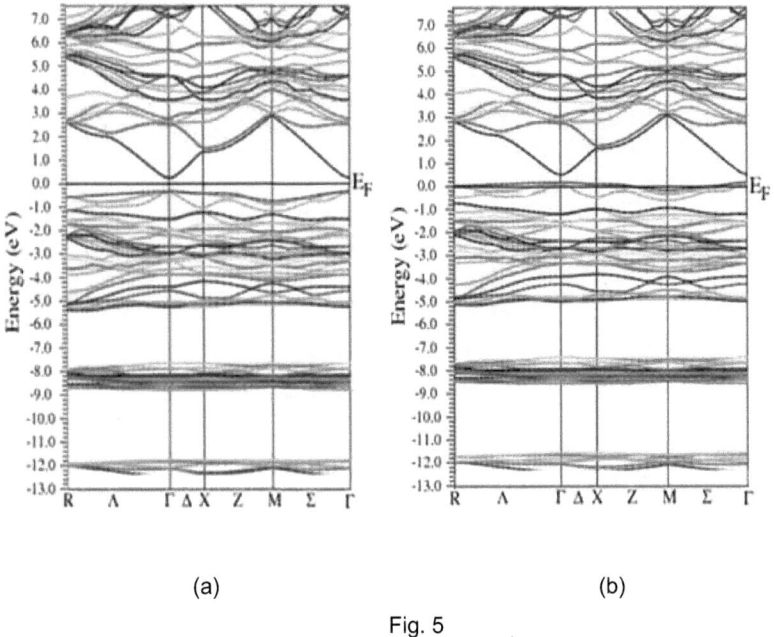

(a)   (b)

Fig. 5

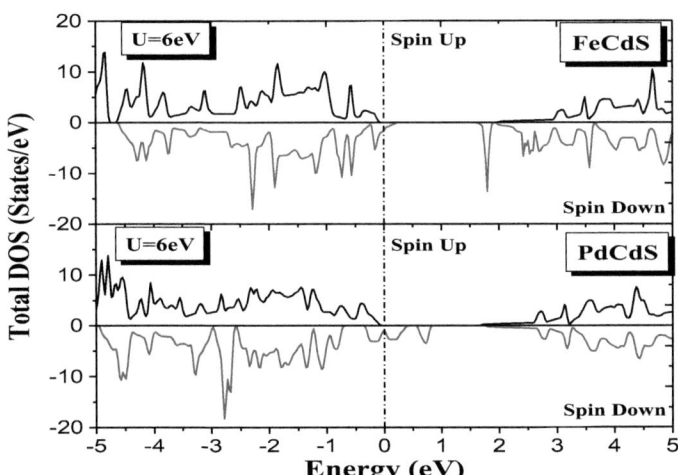

Fig. 6

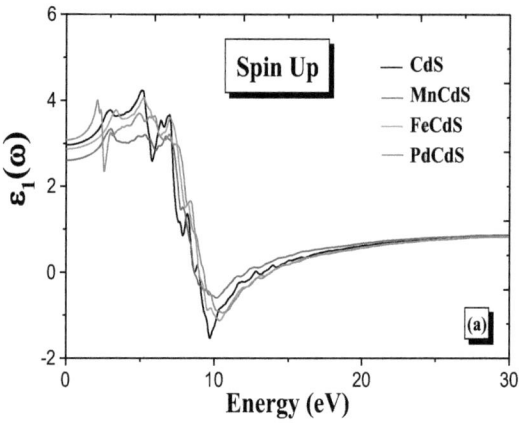

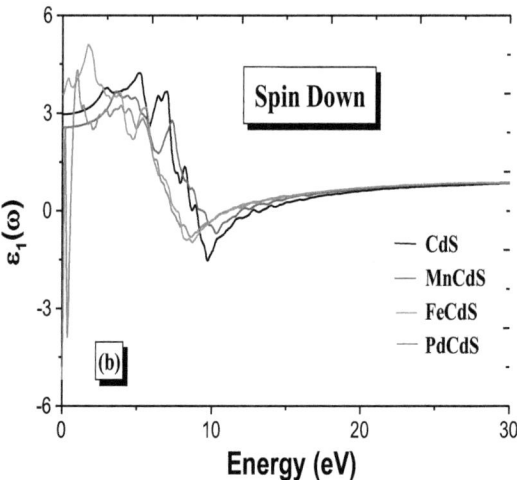

Fig. 7

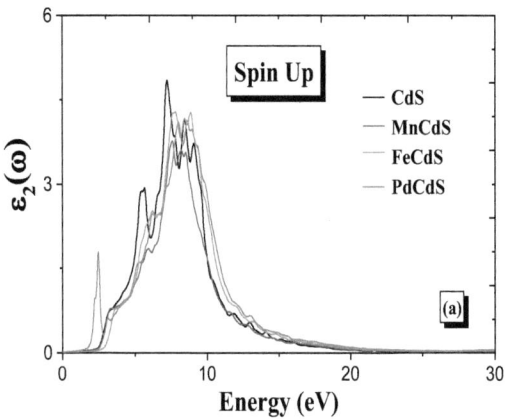

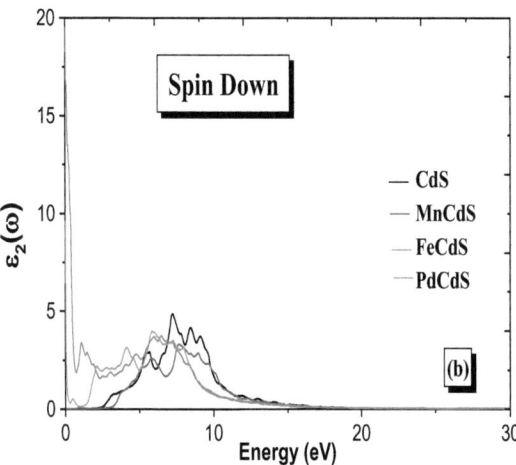

Fig. 8

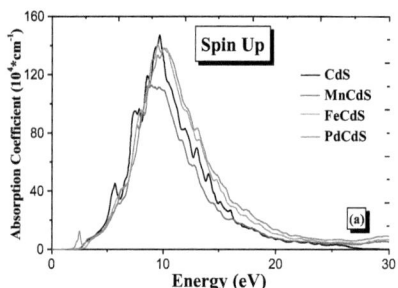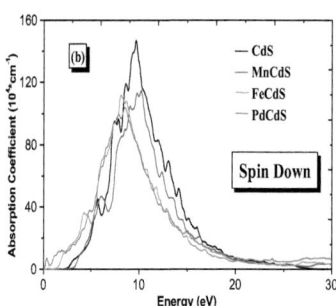

Fig. 9

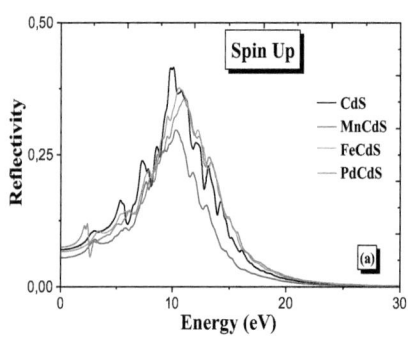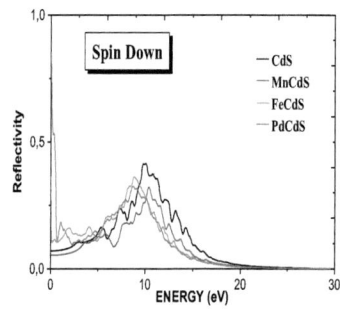

Fig. 10

Site	MnCdS(µB)	FeCdS (µB)	PdCdS(µB)
Mn	3.95	-	-
Fe	-	3.15	-
Pd	-	-	0.79
Cd	0.002	0.01	0.01
S	0.02	0.04	0.08
μ_{total}	5.00	4.00	2.00
$\mu_{interstitial}$	0.86	0.48	0.51

Table. 1

First-Principles Study on electronic and magnetic properties of N mono-doped and (N, Co) co-doped ZnO

A. Abbad*[,1], H.A. Bentounes[2], W. Benstaali[1] and A. Belaidi[3]

[1]Laboratory of Material Valorisation
[2] Signals and Systems Laboratory (LSS)
Faculty of Sciences and Technology, BP227
Abdelhamid Ibn Badis University, Mostaganem (27000) Algeria
[3] Automatic and Systems Analysis Laboratory (LAAS), ENSET, Oran 31000, Algeria

Keywords: density functional theory, p-type, ferromagnetism, magnetic moment

Abstract

Using first principles calculations based on the density functional theory and local spin density approximation, we predict magnetic and electronic properties of N mono-doped and (N-Co) co-doped ZnO for different dopants concentration. The results show that ZnO doped with N concentration of 12.5% is p-type, semi-metallic and ferromagnetic due to the strong hybridization effect between N 2p and O 2p states, with a total magnetic moment of 1μ_B mainly arises from N 2p orbitals. Nevertheless we find a deep and narrow acceptor level, resulting in large acceptor ionization energy of ZnO (N). With increasing N concentration to 25% we find that the impurity energy level is shallow and shifts downward to the direction of low energy, consequentially, the acceptor binding energy is reduced. (N-Co) co-doped ZnO with a concentration of 12.5% for the two dopants is p-type and half-metallic with an important magnetic moment of 3.98 μ_B, due to Co 3d and N 2p states.

1. Introduction:

Diluted magnetic semiconductors (DMS) are semi conducting materials in which a fraction of the host cations can be substitutionally replaced by magnetic ions or appropriate rare earths. Much of the attention on DMS materials is due to its potential application in "spintronics" devices, which exploit spin in magnetic materials along with charge of electrons in semiconductors. Transition metal doped ZnO is a promising candidate material for the field of spintronics. Zinc oxide (ZnO) is an II–VI compound semiconductor with a hexagonal wurtzite structure. It is a promising optoelectronic material that is effective in the UV or blue light region due to its direct

49

wide band gap (3.37 eV) and large exciton binding energy (60 meV) and has been studied by a variety of growth techniques [1-3]. Since Dietl et al predicted that TM doped ZnO might have a Curie temperature above room temperature [4], wide studies have been carried out on ZnO based DMSs and there have been many reports of room temperature ferromagnetism in 3d TM-doped ZnO DMSs[5-10]. The occurrence of ferromagnetism as well as its physical origin in transition metal doped ZnO is controversial. Contradictory reports are available: paramagnetic [11,12], ferromagnetic (FM) [13,14] and spin-glass [15]. Experimental study by Liu et al. [16] as well as first-principles calculations of Zhang et al. [17] shows the stabilization of ferromagnetism in $Zn_{1-x}Co_xO$ when co-doped with Al. Nitrogen has been considered as a suitable acceptor for making p-type ZnO [18-25] that can greatly improve the performance of ZnO-based optoelectronic devices. Recently N-doped ZnO was found to be magnetic even without the presence of any traditional magnetic elements. Yu *et al* [26] prepared N-doped ZnO thin films using pulsed laser deposition and observed ferromagnetic (FM) behavior. In this paper we describe the results of density functional theory (DFT) based first principles calculation for electronic and magnetic properties of N-doped and N-Co co-doped ZnO and obtain some interesting results.

2. Calculation:

The electronic and magnetic properties were studied by using the WIEN2K program [27]. It is based on the full-potential linearized augmented plane wave (FPLAPW) method. The electrons exchange-correlation energy is described in the local spin density approximation (LSDA). The muffin-tin radii (RMT) are 1.95, 1.60, 1.60, and 1.90a.u, for Zn, O, N, and Co, respectively. We use 400 k points in the first Brillouin zone. In order to achieve energy eigenvalues convergence, the wave functions in the interstitial region were expanded in plane waves with a cut-off of kmax =7.5/Rmt (where Rmt is the average radius of the MT spheres). In our article, the electronic and magnetic properties of N doped ZnO at a concentration of 6.25% and 12.5% are calculated using a 2×2×1 supercell containing 16 atoms. In this article, we have first, for each doping concentration, optimized many configurations, but we have used, only the configuration which corresponds to the lowest minimum total energy. For Zn_8NO_7, we have substituted the first O atom which has the atomic position: (0.16; 0.33; 0.87) of the structure by one N atom and for $Zn_8N_2O_6$, the first (0.16; 0.33; 0.87) and the second (0.16; 0.33; 0.37) O atoms are replaced by two N atoms. Other forms of defects such as interstitials or antisites are not considered in this article. In co-doping process, the first Zn atom (0.16; 0.33; 0.5) is substituted by one Co atom.

3. Results and discussions

The structural optimizations are first performed for wurtzite N-doped ZnO super-cell. The optimized lattice parameters are a=3.2028 Å and c =5.1435 Å, which are in agreement with the experimental and other theoretical values [28].For N-doped ZnO, the calculated DOS is shown in Fig. 1(a-d), where the dotted line indicates the Fermi level which is specified to be zero in this paper. The valence band of N-doped ZnO can be divided into three regions: the first one is from −6.5 eV to −4.9 eV with a strong character of d, mostly originating from the 3d state of Zn atoms and a little contribution from p state of O and N can be seen. The second one from -4.9 eV to -4 eV occurs mainly from Zn 3d states. Finally, the upper valence band, originates from a strong hybrid interaction between O 2p and N 2p states which confirms the semi-metallic character of N-doped ZnO. The 2p orbitals of N are spin polarized (see fig. 1(c)), resulting in an asymmetric spin-up and spin-down DOS (fig. 1(a)) and making the system semi-metallic. The O 2p orbitals are also slightly polarized (fig. 1(d)) but as we can see from fig. 1(b) there is no contribution from Zn 3d orbitals to the magnetic moment. Consequently the observed magnetism is primarily from N 2p orbitals. We observe that the energy gap is reduced due to N states above the valence band maximum, suggesting that N-doping can enhance electronic conductivity.

The calculated total magnetic moment in N-doped ZnO is about $1\mu_B$. The magnetic moment on N atom is $0.32\mu_B$. The neighbouring Zn and O atoms are polarized and carry a small moment of about $0.04\mu_B$.

The replacement of O with N leads to holes in N 2p states which create local magnetic moments and result in p-type ZnO. Our results are in good agreement with ones reported by Q.Wang [29]. The peak of impurity states is about 0.1 eV away from the top of the valence band and acts as a deep acceptor level. Holes generated around the top of the valence band are very localized by repulsion effects and we find narrow N-impurity band for N doped ZnO, resulting in large acceptor ionization energy of ZnO (N). ZnO doped with 12.5% of nitrogen has low doping concentration and efficiency. This may possibly be caused by the expelling effect of holes between the valence band and N impurities [30]. From above findings, N-impurity of the material is required to change from narrow bands to broad ones, in order to obtain the reduced acceptor binding energy. We try to solve this problem by increasing N concentration or by co-doping method. The total DOS of ZnO (2N) is shown in fig. 1(e). Compared to the situation of fig. 1(a) the impurity energy level shifts downward to the direction of low energies. The impurity energy level is shallow and locates just at the Fermi level, and the top of the valence band at about -0.11 eV. The total magnetic moment in the cell has increased and is about $1.62\mu_B$. The magnetic moment on each N atom is $0.40\mu_B$, for Zn and O atoms is 0.08 and $0.06\mu_B$ respectively. We can see from fig. 1(e) a change from narrow N-impurity bands for ZnO (N) to broad ones for ZnO (2N) due to the strong hybridization between p states of the two N atoms resulting in the reduced acceptor binding energy. We can also note that the impurity energy level in the total DOS plot shifts toward lower energies region. This means the stabilization of the ionic charge distributions in the material.

In order to investigate the effect of co-doping on electronic and magnetic properties of N doped ZnO, we have replaced one Zn atom in the supercell with one Co atom. The corresponding total and partial DOSs are plotted in fig. 2(a-e). We can see clearly that the material which was semi-metallic becomes half-metallic due to hybridization of the majority Co 3d and N 2p states. Comparing to fig.1 (a), the Co impurities are introduced throughout the valence band. New localized states are seen in the middle of the gap, which correspond to a partial electron doping and are an artefact of the LSDA solution. The impurity energy level locates just at -0.08 eV below Fermi level. From the partial DOS (see fig. 1(b) and 1(d)), the impurity band is due essentially to overlapping between Co 3d and N 2p states. The total magnetic moment in the cell has enhanced and is about 3.98 μ_B with 2.60 μ_B coming from Co atom and 0.37 μ_B from N atom. Zn and O contribute a slightly in the magnetic moment. In table 1 we resume the different values of magnetic moments for each structure. We can conclude that $CoZn_7NO_7$ is p-type with a lower level of acceptor and interesting magnetic moment. Figures 2 (f-j) shows the total and partial DOS of ZnO (2N+CO). From the plots it can be seen that N 2p an O 2p states are shifted in the direction of high energies toward the conduction band and this is an inconvenient because large activation energy is needed to create holes in the valence band. It can be noted again from partial DOS's that localized states become increasingly delocalized, and the total DOS at Fermi level is drastically reduced, which may probably reduces Curie's temperature.

4. Conclusion:

From first principles study, the electronic and magnetic properties of N mono-doped and (N-Co) co-doped ZnO have been investigated. A series of ZnO supercell models were constructed with different concentrations of the dopants. The total and partial density of states were calculated and analyzed. Results show that ZnO doped with 12.5% of Nitrogen is p-type and semi-metallic due to the strong hybrid interaction between O 2p and N 2p states. We find a narrow N-impurity band, resulting in large acceptor ionization energy of ZnO (N). With increasing N concentration to 25%, the impurity energy level is shallow and shifts to the direction of low energy which may reduce the acceptor binding energy and stabilize the ionic charge distributions in the material. The co-doping of Co donors with N-acceptors with a concentration of 12.5% for the two dopants causes an important change from semi-metallic material to half-metallic one, and from localized N-impurity states for ZnO (N) to the delocalized ones for p-type co-doped materials with the enhancement of the total magnetic moment. Finally, it is possible to obtain p-type ZnO with an estimable conductivity and magnetic properties when suitable dopants concentrations are selected. We should notice that, in this paper, we have employed rather small supercells. To further enhance our results, it would be preferable, for future work, to increase greatly the size of the structures used. This will allow us to have more disorder. But to use the Wien2k code in this case, it would be necessary to use before, the concept of special quasirandom structures (SQS's) proposed by Zunger et al [31] which consists in constructing a medium supercell, which periodically repeated, can reproduce the statistical properties of a larger doped structure.

References:

[1] J. Wang, V. Sallet, F. Jomard, A.M Botelho do Rego, E. Elangovan, R. Martins and E. Fortunato, Thin Solid Films. 515 (2007) 8780
[2] X. Guo, H. Tabata and T. Kawai J. Cryst. Growth, 223 (2001) 135
[3] D. C. Look, D. C. Reynolds, C. W Litton ,R.L Jones, D.B. Eason and G.Cantwell .Appl. Phys. Lett. 81 (2002) 1830

[4] T. Dietl, H. Ohno, F. Matsukura, J. Cibert, D. Ferrand, Science. 287 (2000) 1019

[5] J. Shim, T. Hwang, J. Park, S. J. Han, and Y. Jeong, Appl. Phys. Lett.,86(2005) 082503

[6] Y. M. Cho, W. K. Choo, H. Kim, D. Kim, and Y. Ihm, Appl. Phys. Lett., 80 (2002) 3358

[7] S. W. Lim, D. K. Hwang and J. M. Myoung, Solid State Commun., 125 (2003) 231

[8] K. Ueda, H. Tabata, and T. Kawai, Appl. Phys. Lett., vol. 79 (2001) 988

[9] P. Sharma, A. Gupta, K. V. Rao, F. J. Owens, R. Sharma, R. Ahuja, J. M. Osorio-Guillen, and G. A. Gehring, films,Nat. Mater. 2 (2003) 673

[10] J. H. Kim, J. B. Lee, H. Kim, D. Kim, Y. Ihm, and W. K. Choo,, IEEE Trans. Magn., 38 (2002) 2880

[11] S.W. Jung, S.-J. An, G.-C. Yi, C.U. Jung, S.-I. Lee, S. Cho, Appl. Phys. Lett. 78 (2001) 958

[12] X.M. Cheng, C.L. Chien, J. Appl. Phys. 93 (2003) 876

[13] S. Kolesnik, B. Dabrowski, J. Appl. Phys. 96 (2004) 5379

[14] T. Fukumura, Z. Jin, M. Kawasaki, T. Shono, T. Hasegawa, S. Koshihara, H. Koinuma, Appl. Phys. Lett. 78 (2001) 958

[15] A.K. Pradhan, K. Zhang, S. Mohanty, J.B. Dadson, D. Hunter, J. Zhang, D.J. Sellmeyer, U.N. Roy, Y. Chui, A. Burger, S. Matthews, B. Joseph, B.R. Sekhar, B.K. Roul, Appl. Phys. Lett. 86 (2005) 52511

[16] X.C. Liu, E.W. Shi, Z.Z. Chen, H.W. Zhang, B. Xiao, L.X. Song, Appl. Phys. Lett. 88 (2006) 2503

[17] T. Zhang, L.-X. Song, Z.-Z. Chen, E.-W. Shi, L.-X. Chao, H.-W. Zhang, Appl. Phys. Lett. 89 (2006) 172502

[18] L. Dunlop, A. Kursumovic and J.L . MacManus-Driscoll. Appl. Phys. Lett. 93 (2008) 172111

[19] D. C. Oh,J.J Kim ,H. Makino ,T. Hanada ,M.W. Cho ,T. Yao and H.J. Ko, Appl. Phys. Lett. 86 (2005) 042110

[20] Y.J. Zeng ,Z.Z Ye ,Y.F Lu, W.Z. Xu, L.P Zhu ,J.Y. Huang, H.P He and B.H. Zhao, J. Phys. D: Appl. Phys. 41 (2008) 165104

[21] G. D. Yuan, W. J Zhang,J. S. Jie ,X. Fan , J.A Zapien ,Y.H Leung, L.B Luo, P.F Wang , C.S. Lee and s.t. Lee, Nano. Lett. 8 (2008) 2591

[22] Nakano Y, Morikawa T, Ohwaki T and Taga Y, Appl. Phys. Lett. 88 (2006) 172103

[23] E. Kaminska et al, Phys. Status Solidi c 2 (2005) 1119

[24] M. Hirai and A. Kumar, J. Vac. Sci. Technol. A 25 (2007) 1534

[25] L. L Chen, J. G Lu ,Z. Z Ye,Y. M Lin,B. H Zhao,Y. M Ye ,J. S Li and L. P Zhu ,Appl. Phys. Lett. 87 (2005) 252106

[26] C. F Yu ,T. J Lin,S. J Sun and H. Chou , Phys. D: Appl. Phys. 40 (2007) 6497

[27] P. Blaha, K. Schwaz, G. K. H. Madsen, D. Kvasnicka, and J.Luitz, WIEN2K, An Augmented Plane Wave and Local Orbitals Program for Calculating Crystal Properties (Technical University Wien, Vienna, 2001).

[28] H. Rozale, A. Lakdja, A. Lazreg, and P. Ruterana, Phys. Status Solidi B. 7 (2010) 247

[29] Q.Wang, Q.Sun and P.Jena, New Journal Of Physics 11 (2009) 063035

[30] X.Ma, Journal Of Nanomaterials. 10 (2011) 952616

[31] A. Zunger, S. H. Wei, L. G. Ferreira, and J. E. Bernard, Phys.Rev. Lett. 65 (1990) 353

TABLE

Table1: Calculated Magnetic Moments (in Bohr Magneton μB) For Several Sites Of Each Structure

FIGURES CAPTIONS

Fig.1:   (a) Total DOS of Zn_8NO_7, (b) partial DOS of Zn, (c) and (d) partial DOSs of N and O, respectively, (e)Total DOS of $Zn_8N_2O_6$, (f) partial DOS of Zn, (g) and (h) correspond to partial DOSs of N and O, respectively.

Fig.2:   (a) Total DOS of $CoZn_7NO_7$, (b) partial DOS of Co, (c) partial DOS of Zn (d) and (e) partial DOSs of N and O, respectively, (f)Total DOS of $CoZn_7N_2O_6$, (g) partial DOS of Co, (h) partial DOS of Zn, (i) and (j) correspond to partial DOSs of N and O, respectively.

Site	Zn_8NO_7	$Zn_8N_2O_6$	$CoZn_7NO_7$	$CoZn_7N_2O_6$
Zn	0.005-0.04	0.03-0.08	0.01-0.03	0.02
O	0.04-0.05	0.02-0.06	0.02-0.14	0.01-0.13
N	0.32	0.40	0.34	0.25
N_2	-	0.40	-	0.22
Co	-	-	2.60	2.41
μ_{total}	0.92	1.62	3.98	3.80
$\mu_{interstitiel}$	0.11	0.19	0.37	0.41

Table1

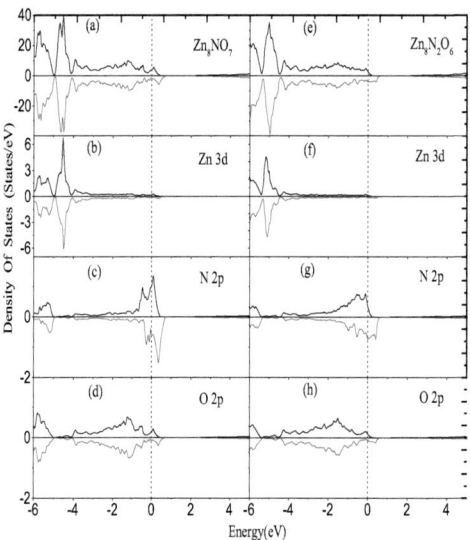

Fig. 1.

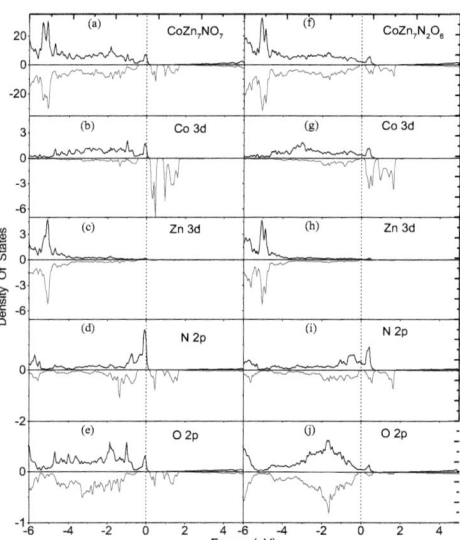

Fig. 2.

Study of electronic and magnetic properties of binary zinc sulfide and ternary manganese -and iron- substituted alloys

A.Abbad[1], S. Bentata[2], H.A. Bentounes[1], W.Benstaali[2] and B.Bouadjemi[2]

[1] Signals and Systems Laboratory (LSS)
Faculty of Sciences and Technology, BP227
Abdelhamid Ibn Badis University, Mostaganem (27000) Algeria
[2] Laboratory of Material Valorisation
Faculty of Sciences and Technology, BP227
Abdelhamid Ibn Badis University, Mostaganem (27000) Algeria

*E-mail: ben_wissam@yahoo.fr

Abstract

A series of first principles calculations have been carried out to investigate magnetic and electronic properties of zinc manganese sulfide ZnMnS for three different manganese concentrations and for zinc iron sulfide ZnFeS. The electronic structures and properties of ZnMnS and ZnFeS were calculated by the local spin density approximation method based on density functional theory. The electronic-state densities in ZnFeS show that the density of states near Fermi energy level is mainly derived from the orbital (d) of the transition metal. Moreover, ZnMnS for a concentration of 6.25% and 12.5% behaves as a semiconductor and has a weak ferromagnetic behavior for a concentration of 18.75%. Consequently, ZnMnS is unsuitable for spintronics. In contrast, ZnFeS is half-metallic and ferromagnetic for low concentration of Fe, so it is a promising candidate for applications in spintronics.

Keywords: LSDA, Ferromagnetism, Magnetic Moments, half-metallic, Covalent character.

1. Introduction:

Zinc Sulfide (ZnS) is a commercially important II–VI semiconductor with a wide optical band gap of 3.68 eV at room temperature [1], making it a very attractive option for optical application especially in nanocrystalline material. ZnS has two different crystal structures (zinc blende and wurtzite). It is considered to be a promising host material for light emitting diode [2, 3], an antireflection coating for heterojunction solar cells [4], and other optoelectronic devices such as electro luminescence devices and photovoltaic cells [5, 6]. ZnS has a low exciton Bohr radius (2.5 nm), that, makes its nanoparticles interesting as small biomolecular probes for fluorescence and laser scanning microscopy. ZnS is also currently used as a shell or capping layer in core/shell nanoprobes such as CdSe/ZnS core/shell structures [7]. Optical and luminescent properties of nanocrystalline ZnS prepared in the forms of thin film, powder and colloid using different synthesis techniques such as sputtering [8], sol-gel [9], solid state [10], micro-wave irradiation [11, 12], ultrasonic or synthesis under high-gravity environment were also studied [13].

In the last years, a number of transition metal (TM) have been introduced in semiconductors at room temperature and a ferromagnetic behavior has been reported [14–16]. Some groups have reported that the intrinsic nonmagnetic elements, such as, Mg [17], Li [18], and Cu [19] as dopants, can order ferromagnetism in some semiconductors hosts and others have reported a non-magnetic behavior [20]. Recently, experimental study of M. El-Hagary [21], showed the absence, at room temperature, of ferromagnetism in all nanocrystalline $Zn_{1-x}Mn_xS$ films (x=0.02, 0.05, 0.1 and 0.15)

Motivated by these findings, we investigated the electronic and magnetic properties of ZnMnS and ZnFeS. We present our results on three different ZnMnS alloys ($Zn_{15}MnS_{16}$, $Zn_{14}Mn_2S_{16}$ and $Zn_{13}Mn_3S_{16}$), and the effect of only one Fe atom introduced into ZnS, corresponding to $Zn_{15}FeS_{16}$.

2. Calculation:

The electronic and magnetic properties were studied by using the WIEN2K code [22]. It is based on the full-potential linearized augmented plane wave method (FPLAPW) [23]. The electrons exchange-correlation energy is described in the local spin density approximation (LSDA), The muffin-tin (MT) radii of Mn,Fe, Zn and S were chosen to be 1.90, 1.90, 2, and 1.8 respectively. Basis functions were expanded as combinations of spherical harmonic functions inside non-overlapping spheres around the atomic sites (MT spheres) and in Fourier series in the interstitial region. We use 200 k points in the first Brillouin zone which were obtained after an energy convergence test (74 k points in the reduce wedge). The wave functions in the interstitial region were expanded in plane waves with a cutoff of kmax=8/RMT (where RMT is the average radius of the MT spheres). The supercell employed contains 32 atoms, which correspond to a 2×2×2 supercell of ZnS. The self-consistent calculations were considered to be converged when the energy convergence is less than 10^{-4} Ry. The lattice parameters which minimize the total energy are (in Å radii) a=3.7366, and c=6.7366 which are in good agreement with other theoretical calculations [24-27] and experimental values [28, 29].

Three different ZnMnS alloys were checked : $Zn_{15}MnS_{16}$, $Zn_{14}Mn_2S_{16}$ and $Zn_{13}Mn_3S_{16}$ respectively (see Figure1), the atomic positions of Mn were chosen in order to have structures with the lowest lattice parameters. We have also studied ZnS with 6.25% and 12.5% of Fe.

3. Results and discussions

The structural properties of ZnMnS and ZnFeS ternary compounds in ferromagnetic (FM) and antiferromagnetic (AFM) phases are determined by calculating the total energy of several different volumes with 6.25%, 12.5% and 18.75% of Mn and 6.25%, 12.5% of Fe. In table 1, we give our calculated results for equilibrium lattice parameters a and c of each compound and the pressure derivative of bulk modulus B'. We have found that with increasing the concentration of Mn or Fe, the values of a and c decrease slightly.

Figure (2-a) illustrates the total electronic density of states (DOS) of pure ZnS supercell. The calculated results show that pure ZnS has a direct band gap of about 2.04 eV, which is smaller than the experimental value (3.68 eV), because the LSDA underestimates the band-gap. The DOS of the majority spin and the DOS of the minority spin are symmetric and there is no resultant spin polarization. In figures (2-b), (2-c) and (2-d) we plot the total electronic density of states (DOS) of $Zn_{15}MnS_{16}$, $Zn_{14}Mn_2S_{16}$ and $Zn_{13}Mn_3S_{16}$ respectively. We can see from figure (2-b) that the Fermi level is situated through the band gap, this means that the compound keeps its semi conducting behavior, we can also observe a reduction in the band gap compared to pure ZnS, this is due to the appearance of localised states in the minority spin of the conduction band (CB), which move toward the low energies, and in the majority spin of the valence band (VB) which move slightly toward the high energies. Consequently, we can say that the influence of one Mn atom on the electronic properties of $Zn_{15}MnS_{16}$ is very low. Figure (2-c) shows the total DOS of $Zn_{14}Mn_2S_{16}$, we can see that, even though the energy gap narrows considerably, the Fermi level still lies in a region of vanishing electron density. So, we can say that $Zn_{14}Mn_2S_{16}$ behaves also as a semiconductor despite that the concentration is increased to 12.5% and despite the fact that the total magnetic moment becomes 9.77 µB (see table 2) which is a high value. We have also calculated the energy difference ($\Delta E = E_{AFM} - E_{FM}$) between the ferromagnetic (FM) and antiferromagnetic (AFM) states for this alloy (table 1) which is equal to -24 meV, this means that Mn atoms are coupled antiferromagnetically because E_{AFM} is lower than E_{FM}.

We have then increased the Mn concentration until 18.75% (Figure 2-d). In this case, we note that the impurity bands are located within the band gap and that the Fermi level falls in the gap within the impurity bands of the spin-down electrons. The difference in the calculated total energies of the system between the (FM) and (AFM) states is 0.07meV with the ferromagnetic state being the ground state because E_{FM} is lower than E_{AFM}. Although the fact that the ferromagnetic coupling is confirmed by this positive energy ΔE, it is too much small to stabilize ferromagnetism at room temperature.

To understand the nature of the electrons at the Fermi energy, we have calculated the partial density of states around Mn, S and Zn atoms in Fig 3(a, b, c). In figures 3(a) and 3(b), we remark that neither Mn nor S introduce DOS at Fermi energy although there is an overlap between Mn (3d) and S (2p) states. In figure 3(c), we can see distinct hybridization between Mn (3d) and S (2p) states in the spin-down bands which leads to an important DOS at the Fermi level.

These results indicate that the Mn concentration changes slightly the electronic properties of semi-conducting ZnS, and we can conclude that with increasing the number of Mn atoms in pure ZnS, we observe a slight change in the behavior of the new compound and we can consequently say that ZnMnS is unsuitable for spintronics. Additionally, ZnMnS alloys start to be ferromagnetic only upper to 20% of Mn atoms which may alter their semiconducting properties.

The magnetic properties of the three alloys were computed. The total magnetic moments for $Zn_{15}MnS_{16}$, $Zn_{14}Mn_2S_{16}$ and $Zn_{13}Mn_3S_{16}$ respectively, are mainly contributed by the Mn atoms. We can see from table 2 that each Mn atoms carry a spin moment of about 3.75µB and that Zn and S atoms also contribute to magnetic moment but with a small part due to hybridization.

In figure (4), we plot the variation of the band gap of ZnMnS as function of Mn variations. Our results show that the band gap of ZnMnS decreases considerably with increasing Mn concentration, and this is mainly due to the move of the minority spin states of the CB toward the Fermi level.

To see if all transition metals have the same effect of Mn on the magnetic and electronic properties on pure ZnS, we have chosen to study the influence of an another transition metal. One of the zinc atoms in the supercell is substituted by one Fe atom which corresponds to $Zn_{15}FeS_{16}$ compound. The DOS of the majority spin and the DOS of the minority spin are significantly modified around the Fermi level, as shown in Fig 2(e). Just below the Fermi level, there is a peak in the minority spin channel while there are no states in the majority spin. However, some localized unoccupied states appear above the Fermi level in the minority spin. We can also deduce that ZnFeS is half metallic since the Fermi energy passes through the DOS of spin down only. Figure 3(d) shows the partial density of states of ZnFeS compound. The substitution of Zn by Fe does not lead to the spin polarization of the conduction band but results in a significant spin polarization of the valence band and a strong coupling between the 2p orbitals of S atoms and the 3d orbitals of Fe atoms near the Fermi level can be seen. The magnetic moment is 3.00 μB per Fe atom (table 2). The magnetic moment is mainly contributed by the iron 3d orbitals. The neighboring Zn and S atoms are also spin polarized and the magnetic moments, 0.01 μB for Zn atoms and 0.06 μB for S atoms, respectively, are positive (parallel to the moment of Fe).The calculated total energy differences ΔE between (FM) and (AFM) states for $Zn_{14}Fe_2S_{16}$ is 0.9 meV, indicating that ferromagnetic coupling is favorable for ZnFeS which is in good agreement with other experimental and theoretical studies [30]. In opposition to ZnMnS, the electronic and magnetic properties of binary ZnS, when only one Fe atom is introduced, change clearly, so we can conclude that ZnFeS is a promising candidate for applications in spintronics.

For an insight into the magnetism in ZnFeS, our calculations show that the magnetic moment of $FeZn_{15}S_{16}$ is important due to Fe as we can observe from table 2, also the magnetic moment of Fe is smaller than the magnetic moment of Mn for all the 3 cases but despite this, ZnFeS is ferromagnetic.

To visualize the nature of the bound character and to explain the charge transfer and the bonding properties of ZnMnS and ZnFeS, we have explored the effect of Mn and Fe states on the charge spin densities. Fig. 5, 6, and 7 show the charge spin densities for pure ZnS, Spin-up and spin-down for ZnMnS and ZnFeS. Since, there is a large difference of the electro-negativity between Zn and S, we observe on Fig. 5 a transfer of charge from Zn to S atoms. Zn and S atoms shared electron that causes the strong covalent interaction of the Zn-S bonds in ZnS. Therefore ZnS has strongly covalent and partially ionic bond simultaneously. From figures 6 and 7 we can see that there are a strong covalent character and a partial ionic bond connecting Mn-S and Fe-S atoms. We can deduce that Mn and Fe haven't significant effect on the nature of the bound character of ZnS.

4. Conclusion:

We have presented in this paper the electronic and magnetic properties of ZnS with 6.25, 12.5, and 18.75% of manganese using the LSDA (Local Spin Density Approximation) method based on density functional theory (DFT). It is pointed out that the Mn concentration can slightly change the electronic properties of semiconducting ZnS which become half-metallic due to partial hole doping of the material at a concentration upper to 20%. The total magnetic moment increases considerably with increasing Mn concentration, evidently the band gap is clearly reduced. These results suggest that MnZnS may be an unsuitable material for applications in spintronics. On the contrary, we found that FeZnS is a half metallic and ferromagnetic compound for a low concentration of Fe (6.25%). Consequently, FeZnS has great potential for successful implementation into spintronic devices. This study shows that the two transition metals used give contradictory results and therefore it would be useful for future work to study the effect of other TM on the behavior of ZnS and to compare it with this present work. It would be also very interesting to study the effect of temperature on the ferromagnetic character of the different compounds.

References:

[1] K. Sooklal, B.S. Cullumn, S.M. Angel, C.J. Murphy, J. Phys. Chem. 100 (1996) 4551-4555.
[2] H. Katayama, S. Oda, H. Kukimoto, Appl. Phys. Lett. 27, (1975), 657.

[3] A. Antony, K.V. Mirali, R. Manoj, M.K. Jayaraj, Mater.Chem.Phys. 90 (2005) 105-106.

[4] W.H. Bloss, F. Pfisterer, H.W.Schock " Advances in solar energy , an Annual review of research & development ". Vol.4 (1988) P-275.

[5] M.C. Beard, G.M. Turner, C.A. Schmuttenmaer, Nano Lett. 2 (2002) 983.

[6] R.P. Raffaelle, S..L. Castro, A.F. Hepp, S.G. Bailey, Prog. Photovoltaics. 10 (2002) 433- 439.

[7] A. Thakur, C. Fradin, Can. Undergraduate Phys. J. 3 (2005) 7-12.

[8] S.K. Mandal, S. Chaudhuri, A.K. Pal, Thin Solid Films. 30 (1992) 209-213.

[9] B. Bhattacharjee, D. Ganguli, S. Chaudhuri, A.K. Pal, Mater. Chem. Phys.78 (2002) 372.

[10] H.-Y. Lu, S.-Y. Chu, S.-S. Tan, J. Cryst. Growth 269 (2004) 385-391.

[11] Y. Zhao, J.-M. Hong, J.-J. Zhu, J. Cryst. Growth 270 (2004) 438-445.

[12] Y. Ni, G. Yin, J. Hong, Z. Xu, Mater. Res. Bull. 39 (2004) 1967-1972.

[13] J. Chen, Y. Li, Y. Wong, J. Wang, J. Yun, D. Cao, Mater. Res. Bull. 39 (2004) 185.

[14] H. Saito, V. Zayets, S. Yamagata, and K. Ando, Phys. Rev. Lett. 90 (2003) 207202_1-207202_4.

[15] Z. Wang, J. Tang, Y. Chen, et al., J. Appl. Phys. 95 (2004) 7384-7386.

[16] N. H. Hong, J. Sakai, and W. Prellier, Phys. Rev. B 70 (2004) 195204.

[17] C. H. Zhang and S. S. Yan, Appl. Phys. Lett. 95 (2009) 232108.

[18] J. B. Yi, C. C. Lim, G. Z. Xing, et al., Phys. Rev. Lett. 104 (2010) 137201_1-137201_4.

[19] J. H. Lee, I. H. Choi, S. Shin,et al., Appl. Phys. Lett. 90 (2007) 032504_1-032504_3.

[20] I. Sarkar, M. K. Sanyal, S. Kar, S. Biswas, S. Banerjee, S. Chaudhuri, S. Takeyama, H. Mino, and F. Komori. Phys. Rev. B 75 (2007) 224409_1-224409_5.

[21] M. El-Hagary and S. Soltan. Solid State Communications 155 (2013) 29-30.

[22] P. Blaha, K. Schwarz, G. K. H.Madsen, D. Kvasnicka, and J. Luitz, WIEN2k, an Augmented Plane WaveRLocal Orbitals Program for Calculating Crystal Properties (Technische Universita ¨t Wien, Austria, 2001).

[23] D. J. Singh, Plane Waves, Pseudopotentials and the LAPW Method (Kluwer Academic Publishers, Boston, 1994).

[24] M. Bilge, S. Özdemir Kart, H.H. Kart , T. Cagin. J.Achi.Mat.Man.Eng. 31 (2008) 29-34.

[25] C.Y. Yeh, Z.W. Lu, S. Froyen, A. Zunger, Zinc-blende-wurtzite polytypism in semiconductors, Physics Review B. 42 (1992) 10086-10096.

[26] E.C. Hu, L.L. Sun, Y.Z. Zeng, R.X. Chen. Chinese Physics Letters. 25 (2008) 675-678.

[27] M. Catti, Y. Noel, R. Dovesi. Journal of Physics and Chemistry of Solids. 64 (2003) 2183-2190.

[28] Y.N Xu, W.Y. Ching. Physics Review B. 48 (1993) 4335-4355.

[29] R.R. Reeber, G.W. Powell. Journal of Applied Physics. 38 (1967) 1531-1538.

[30] F.Zhu, S. Dong, G. Yang. Optoelectronics and Advanced Materials 4 (2010) 2072-2075

TABLES

Table1: The calculated structural constants for binary ZnS and ternary ZnMnS, ZnFeS compounds.

Table2: Magnetic Moments (in Bohr Magneton µB) of Several Sites and the Total Magnetic Moment for Each Structure

FIGURES CAPTIONS

Fig.1: The (2x2x2) supercell of ZnMnS (a) $Zn_{15}MnS_{16}$ (b) $Zn_{14}Mn_2S_{16}$ (c) $Zn_{13}Mn_3S_{16}$

Fig.2: Total Density of States (a) $Zn_{16}S_{16}$ (b) $Zn_{15}MnS_{16}$ (c) $Zn_{14}Mn_2S_{16}$ (d) $Zn_{13}Mn_3S_{16}$ (e) $Zn_{15}FeS_{16}$

Fig.3: Partial Density of States of: (a) $Zn_{15}MnS_{16}$: Mn (d), Zn (d) and S (p) (b) $Zn_{14}Mn_2S_{16}$: Mn (d), Zn (d) and S (p). (c) $Zn_{13}Mn_3S_{16}$: Mn (d), Zn (d) and S (p) (d) $Zn_{15}FeS_{16}$: Fe (d), Zn (d) and S (p)

Fig.4: Variation of the band gap as function of Mn concentration

Fig.5: Electron charge density for pure ZnS

Fig.6: Electron charge density for ternary ZnMnS for: (a) spin-up and (b) spin-down

Fig.7: Electron charge density for ternary ZnFeS for: (a) spin-up and (b) spin-down

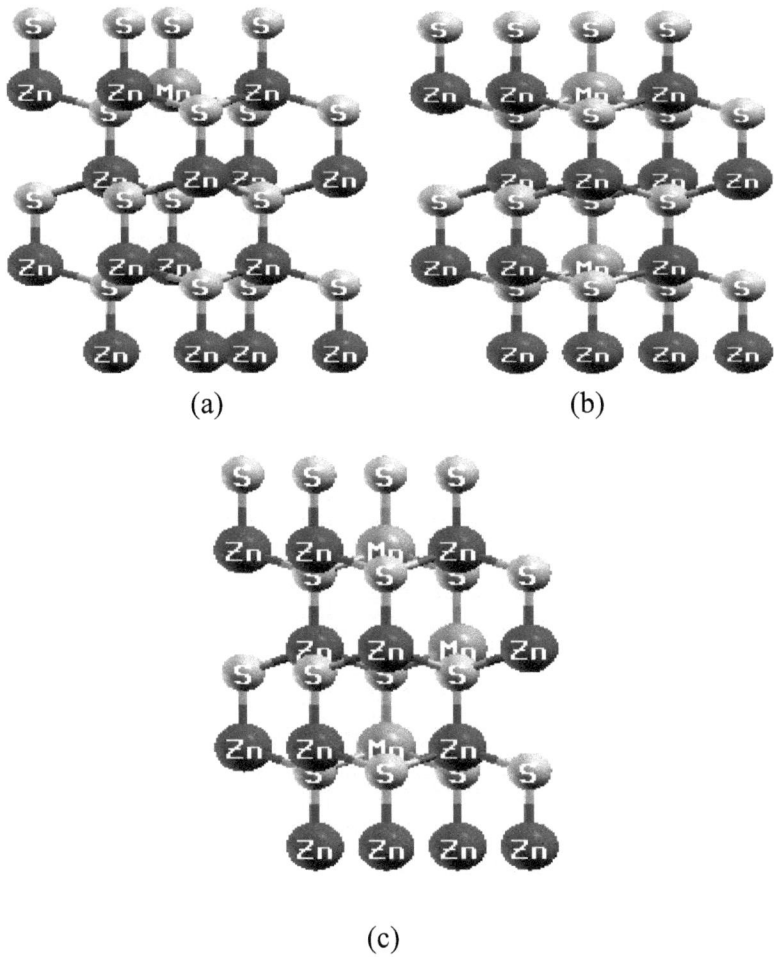

(a)   (b)

(c)

Fig. 1.

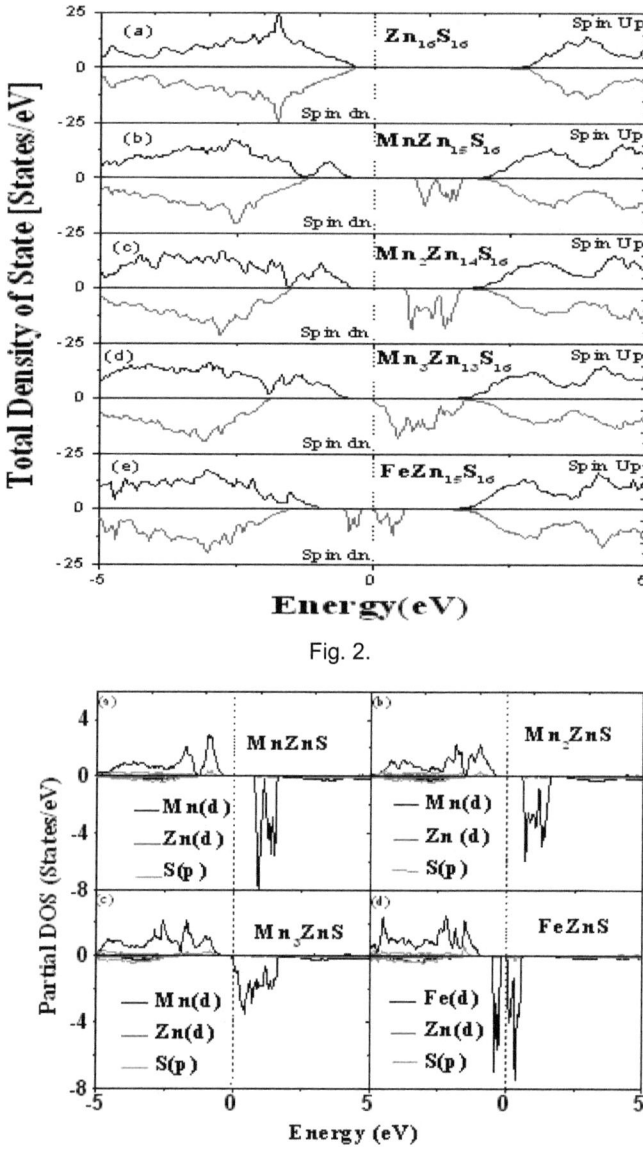

Fig. 2.

Fig. 3.

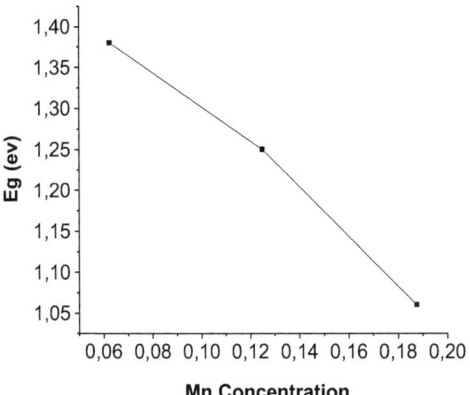

Fig. 4.

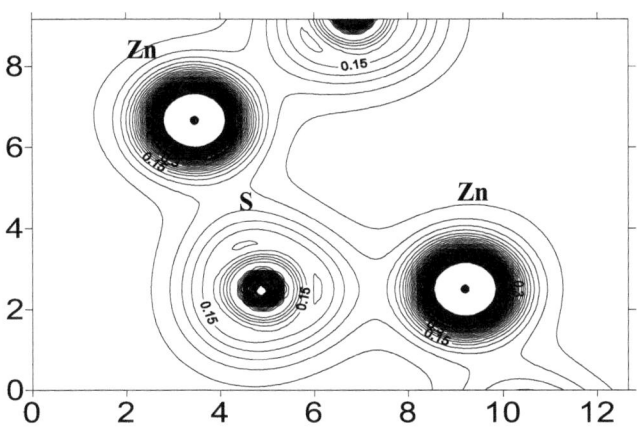

Fig. 5.

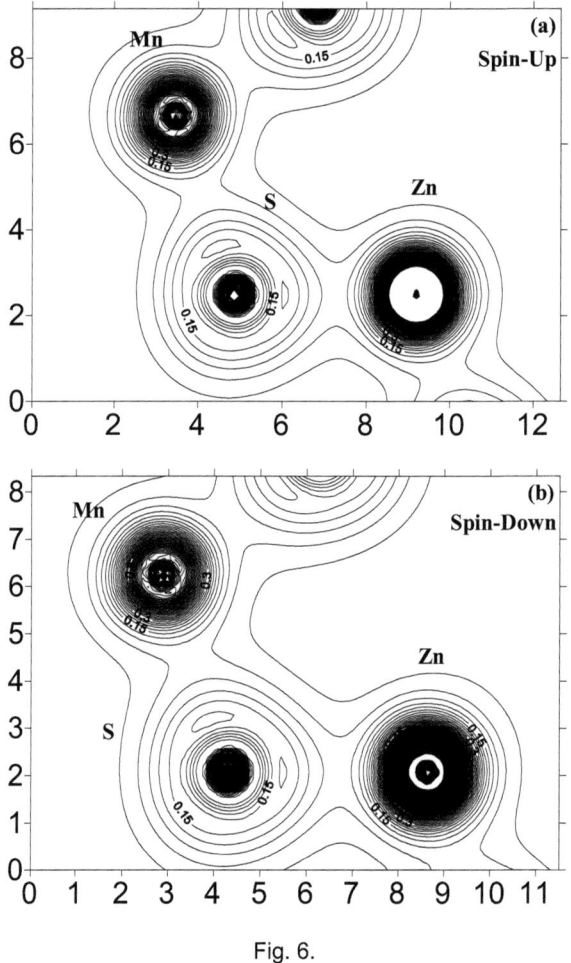

Fig. 6.

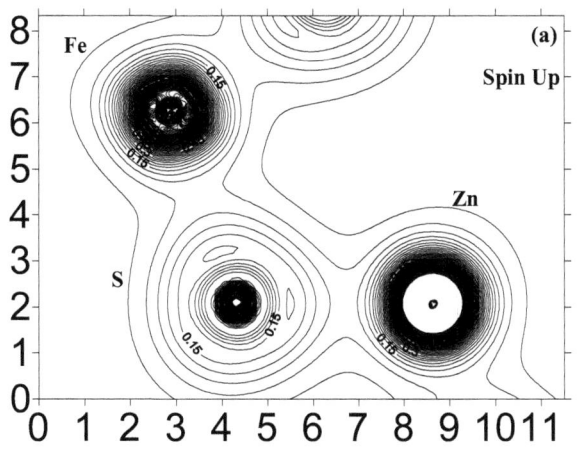

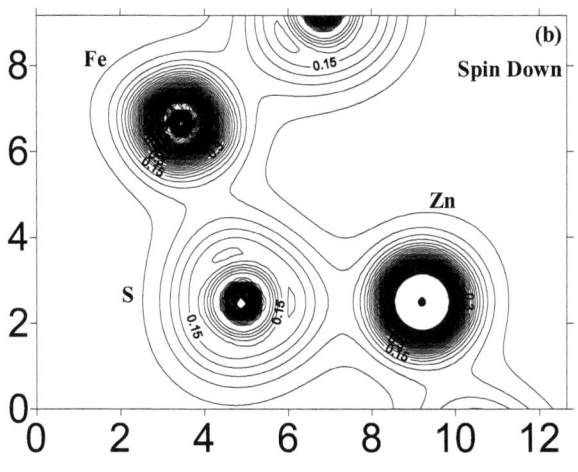

Fig. 7.

Compound	a (A)	c (A)	B'	$\Delta E = E_{AFM} - E_{FM}$ (meV)
ZnS	3.7366	6.7366	4.53	-
$Zn_{15}MnS_{16}$	3.7296	6.7303	4.61	-
$Zn_{14}Mn_2S_{16}$	3.7252	6.7247	4.68	-24
$Zn_{13}Mn_3S_{16}$	3.7214	6.7201	4.82	0.07
$Zn_{15}FeS_{16}$	3.7321	6.7323	4.93	-
$Zn_{14}Fe_2S_{16}$	3.7306	6.7298	5.02	0.9

Table 1

Site	$MnZn_{15}S_{16}$	$Mn_2Zn_{14}S_{16}$	$Mn_3Zn_{13}S_{16}$	$Fe\,Zn_{15}\,S_{16}$
Mn_1	3.75	3.74	3.79	-
Mn_2	-	3.74	3.76	-
Mn_3	-	-	3.79	-
Zn	0.0001-0.01	0.007-0.01	0.009-0.01	0.005-0.011
S	0.005-0.04	0.009-0.05	0.006-0.08	0.001-0.064
Fe	-	-	-	3.00
μ_{total}	4.96	9.77	15.00	4.00
$\mu_{interstitial}$	0.90	1.66	2.76	0.65

Table 2

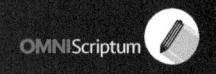

Printed by Books on Demand GmbH, Norderstedt / Germany